Eight Amazing Engineering Stories
Using the Elements to Create Extraordinary Technologies

Also by Bill Hammack

Why Engineers Need to Grow a Long Tail • 2011

How Engineers Create the World • 2011

Eight Amazing Engineering Stories
Using the Elements to Create Extraordinary Technologies

Bill Hammack, Patrick Ryan, & Nick Ziech

Copyright © 2012 William S. Hammack, Patrick M. Ryan, & Nicholas E. Ziech
All rights reserved. No part of this book may be reproduced in any form by any electronic or mechanical means (including photocopying, recording, or information storage and retrieval) without permission in writing from the publisher.

Articulate Noise Books

First Edition: April 2012

    Hammack, William S., Patrick M. Ryan, & Nicholas E. Ziech

    Eight Amazing Engineering Stories: Using the Elements to Create Extraordinary Technologies / Bill Hammack - 1st edition (version 1.4)

    ISBN 978-0-9839661-3-5 (pbk)

    ISBN 978-0-9839661-4-2 (electronic)

    ISBN 978-0-9839661-5-9 (hbk)

    1. Engineering. 2. Technology --Popular works. 3. Chemical elements. 4. Electronic apparatus and appliances. 5. Engineering design I. Title.

Table of Contents

Introduction *1*

Silicon *3*

    Digital Cameras: How a CCD Works *5*

    How a Smartphone Knows Up from Down *23*
      In Depth: The Mathematics of Capacitors 35

Cesium *39*

    How an Atomic Clock Works *41*

Primer: Nuclear Structure *57*

Uranium *63*

    The Hardest Step in Making a Nuclear Bomb *65*

Lead *81*

    The Lead-Acid Battery *83*
      In Depth: Entropy 104

Aluminum *109*

    Anodizing, or The Beauty of Corrosion *111*

Primer: Waves *125*

Tungsten, Thorium, & Copper *135*

    How a Microwave Oven Works *139*

Primer: Electrons, Energy Levels, and Light Emission *165*

Chromium, Helium, & Neon *171*

    How a Laser Works *175*
      In Depth: Semiconductors, Electrons & Holes 197

Detailed Table of Contents

Silicon 3

Digital Cameras: How a CCD Works 5

How a Single Pixel Measures Light Intensity 6
Discovery of the Photoelectric Effect 7
The Photoelectric Effect 8
How Not to Make a Digital Camera 9
Capacitive Coupling 11
CCD: Charge Coupled Device 12
How a CCD Creates Color 18
Active Pixel Sensors (APS) 19
The CCD and the Nobel Prize 21

How a Smartphone Knows Up from Down 23

Basics of an Accelerometer 23
Technology & Privacy 26
The Accelerometer Inside a Smartphone 27
Using a Capacitor to Make an Accelerometer 28
How to Make an Accelerometer 31
In Depth: The Mathematics of Capacitors 35

Cesium 39

How an Atomic Clock Works 41

Basics of Modern Timekeeping 42
Quartz Resonators 44
Cesium-based Atomic Clocks 46
Uses of Accurate Time 50
Throttling a GPS Device 50
How the GPS System Works 51
The Mathematics of GPS 52

Primer: Nuclear Structure 57
- Fission Fission 58
- Nuclear Explosions & Chain Reactions 60

Uranium 63

The Hardest Step in Making a Nuclear Bomb 65
- Operation Alsos: How We Know What Heisenberg Said 66
- The Power of the First Nuclear Bomb 68
- How Uranium Was Enriched for the First Atomic Bomb 69
- Gaseous Diffusion 70
- Modern Method: Centrifuge 71
- Why It's Hard to Make a Rotor 74
- The Future 78

Lead 81

The Lead-Acid Battery 83
- Are Electric Cars Really Better for the Environment? 84
- Basis of All Batteries: Transfer of Electrons 85
- Elements of a Battery 87
- Electrolytes 87
- Engineering a Useful Battery 87
- The First Battery? 90
- Why Discharged Car Batteries Can Freeze 90
- Anodes and Cathodes 92
- Why the Perfect Battery Doesn't Exist 92
- Shallow Charge vs. Deep Discharge 94
- The First Lead-Acid Battery 96
- Why Does the Lead-Acid Battery Still Exist? 98
- How Lithium-Ion Batteries Work 100
- Why Lithium Laptop Batteries Explode 102
- In Depth: Entropy 104

Aluminum 109

Anodizing, or The Beauty of Corrosion 111
What Is Unibody Design? 111
Why Do Metals Corrode? 114
Oxidation-Reduction Reactions 115
How to Control Corrosion 117
Why Stainless Steel Doesn't Rust 120
Anodizing Aluminum 120
Anodizing Titanium 123

Primer: Waves 125
Basic Definitions 125
Interaction of Waves 128
Electromagnetic Waves 132

Tungsten, Thorium, & Copper 135

How a Microwave Oven Works 139
What is a Vacuum Tube? 139
How Microwaves Heat Food 143
Does a Microwave Really Heat from the Inside Out? 145
The Magnetron: The Powerhouse of a Microwave Oven 147
How to Create Electromagnetic Waves 147
Why a Microwave Oven Heats with 2,450 MHz Radiation 149
Design and Operation of the Magnetron 153
How Heating Tungsten Produces Electrons 156
Why Do the Cavities Resonate? 158

Primer: Electrons, Energy Levels, and Light Emission 165

Chromium, Helium, & Neon 171

How a Laser Works 175
    Key Characteristics of a Laser Beam 177
    Why Does Anything Give off Light? 178
    Lasers & Spontaneous Emission 179
    Why Does Stimulated Emission Occur? 182
    The Ruby Laser 183
    Who Really Invented the Laser? 188
    Resonator Cavity 189
    Semiconductor Lasers 192
    Erbium Amplifier for Fiber Optics Cables 194
    In Depth: Semiconductors, Electrons & Holes 197

Preface

After creating our third series of EngineerGuy videos in the spring of 2011, we started getting a lot of requests. Specifically, a lot of people wanted to know even more about the topics we covered in the videos. The problem is that you can only put so much information into a five-minute video. So we wrote this book to give a more complete treatment to the subjects we cover in our fourth series of videos. In this book, we aim to educate the reader on how engineers use elements to create the world around us and to tell the fascinating stories behind a lot of the technology we see today.

Although this book covers some scientifically heavy topics, we do it in a way that any determined reader can understand. To aid the reader we have added several primers on subjects before certain chapters. For instance, before our discussion on the preparation of uranium for an atomic bomb, we have written a short primer on nuclear structure. If you feel familiar with how atoms are composed and what holds them together, feel free to skip this primer and go right to the chapter. However, if you feel you know very little about nuclear structure, this section will provide you with the foundation you need to understand the chapter that follows.

The video series was developed using generous funding from the *Special Grant Program in the Chemical Sciences* from The Camille & Henry Dreyfus Foundation.

While the writing and preparation of this book involved hard work and much exhausting labor, we found it a pleasure to work together to create something that, we hope, will delight and inform our readers.

W.S.H. P.M.R. N.E.Z.

March 2012

Urbana, Illinois

Introduction

In this book we focus on how engineers use the chemical and physical properties of specific elements to create the technological objects that surround us. In eight chapters, we cover eight uses of the elements: CCD imagers, tiny accelerometers, atomic clocks, fissile material for bombs and fuel, batteries, anodized metal surfaces, microwave ovens, and lasers.

We have chosen to discuss these uses of the elements because they highlight the different applications that elements have in our lives. We hope that through these chapters you gain not only a deeper understanding for how these elements are used but also an appreciation for the magnificence of the innovation and engineering that go into the world around us.

After this book was written we used it as the basis for a video series. We recommend that our readers find these short videos—none longer than five minutes or so—at *www.engineerguy.com*. Although this book stands on its own the videos greatly help explain the information we share in this book by showing the principles in action.

Silicon

[sil-i-kon] | Symbol: Si
Atomic number: 14 | Atomic weight: 28.0855 amu

The long and significant impact of silicon on humankind can be seen from the name's etymology. The root of the word comes from the Latin silex, meaning flint, because humans have been using silicon in the form of flint since prehistoric times. Today, of course, no single element better defines our age than silicon: The element lies at the core of every one of our electronic devices. Although it doesn't conduct electricity as well as a metal, it does perform better than, say, a piece of rubber, but more importantly silicon can be made to switch internally with no moving parts between these two extremes of conductivity. Thus creating a switch that controls the flow of electrons in circuit boards. Having a blue-gray color and nearly metallic sheen, silicon is never found pure in nature, but when combined with oxygen, it forms the most abundant compound in the earth's crust: SiO_2, commonly found in sand.

In the next two chapters, we will look at two aspects of silicon. First, we will examine how its electronic properties allow engineers to make the CCD (charge coupled device) that captures images in a digital camera. In the chapter after that, we will look at how the structural properties of silicon allow engineers to create amazing, intricate structures on a tiny scale, such as parts in smartphones. These commonplace items perfectly highlight how necessary silicon is to our world today.

Digital Cameras: How a CCD Works

I HAVE SNAPPED PHOTOGRAPHS around the world. As you might guess, of all the stuff I've seen, I'm fascinated most by engineered things. I've photographed the soaring spire of the Eiffel Tower, the concrete dome of the Pantheon, ancient salt works on the border of Croatia and Slovenia, and the simple yet functional construction of huts in a Masai village in Tanzania. Yet, I find the cameras I use to take these pictures every bit as amazing as the structures themselves.

A camera both captures and records an image. In the earliest cameras, film was able to perform both tasks simultaneously. In traditional film cameras, a lens focused light onto film that was composed of a piece of plastic covered with small grains of light-sensitive silver bromide. In the spots where light would strike, silver ions in the grains changed to an altogether different compound, metallic silver. The more intense the light, the more silver was created in a grain. The image remained latent on the film until the photographer used a chemical process to grow the silver spots until they could be seen by the human eye. The silver areas appear dark in the negative and correspond to the bright parts of the image; different shades of gray depend on how much silver was created in a grain by the intensity of the light.

In today's digital camera, these two functions are split: A light-sensitive CCD (charge coupled device) captures the image and then transfers it to the camera's electronics, which record it. At some level you already understand the concept: If you enlarge a digital photograph on your computer, you can easily see the tiny picture elements (the source of the word *pixel*) that make up the image. It's easy to then understand that each of these corresponds to a section of that CCD, but how exactly does it work?

The key to understanding why digital photography became cheap and ubiquitous lies in appreciating the ingenious way that the CCD transfers the image within the camera. To make this technology easier to understand, we'll start the story of the CCD with a single pixel and then work our way out to how the pixels are linked together.

How a Single Pixel Measures Light Intensity

Since the late nineteenth century, engineers and scientists have known that certain solids will produce current when exposed to light. For example, take a bar of selenium (a dense, purple-gray solid extracted from copper sulfate ores) and attach a wire to each end. Hook an ammeter (which measures current) to the bar and then shine bright light on it: The ammeter's needle will jump because the selenium can change the incident light into a flow of electrons. The silicon used in a CCD exhibits a related behavior, as light causes charges to build up on the surface of the silicon.

At its simplest, this defines how a pixel works: A small section of silicon—typically a little less than 10 square microns (a micron is one millionth of a meter)—in a consumer camera generates a build-up of electrons after being exposed to light. The number of electrons trapped is proportional to the intensity of the light, a phenomenon known as the photoelectric effect.

Eight Amazing Engineering Stories

Discovery of the Photoelectric Effect

In 1873, Willoughby Smith, an engineer for the Gutta Percha Company, tested a new insulating material for submarine cables to transmit telegraph signals, which traveled at the blistering speed of 13 words per minute.

We think of an interconnected and wired world as being a late twentieth century phenomenon, but it really began in the nineteenth. Engineers like Smith connected continents by laying miles of cable on the ocean floor. By 1870, for example, one could send a telegraph all the way from Mumbai to London via submarine cables. In order to work, the cables could not have any electrical faults. For example, if the insulation (which was made of gutta-percha, a natural latex of sorts) on the cable failed, the cable would come into contact with salt water. This shorted the cable, causing the electrical telegraph signals to dissipate into the ocean. So, it was of the utmost importance to ensure the insulation stayed intact as crews laid the cable out from ships into the ocean. As they unrolled the cable from giant spools on deck, they constantly monitored the amount of current flowing from the cable into the ocean. If this number jumped, they hauled the cable back in and quickly fixed the insulation. To make this measurement, the end of the cable on shore had to have a high resistance. Smith did this at first by using a bar of high-resistance selenium to electrically isolate the cable. At first all seemed fine. The "early experiments," he noted, cast "very favorable light for the purpose required." However, he soon noted "a great discrepancy in the tests, and seldom did different operators obtain the same results." He discovered that the odd results came from boxes with sliding covers: When the cover was off, the resistance of the selenium dropped. He then noticed that "by passing the hand before an ordinary gas burner," and thus decreasing the amount of light striking the selenium, he could change the resistance by 15%. Smith had discovered photoconductivity, one of several manifestations of the photoelectric effect; that is, increases in the conductivity of a solid material induced by light.

> **The Photoelectric Effect**
>
> Observed in the nineteenth century, the phenomenon of the photoelectric effect was only understood years later in the early twentieth century. In one of his earliest scientific papers, Albert Einstein was the first to explain the effect. Having read of experiments documenting the ejection of electrons from a metal after being irradiated with ultraviolet light, Einstein postulated that light existed both as a particle and as a wave. He based his theory on three unexplained characteristics of the effect:
>
> 1. Metals will not eject electrons, regardless of the light intensity, unless the frequency of the light exceeds a threshold unique to each type of metal.
>
> 2. The kinetic energy of the ejected electrons is linearly proportional to the frequency of the incident light.
>
> 3. Low intensity light is able to eject electrons from metals, as long as its frequency is above the threshold of the metal.
>
> This suggested to him that the photoelectric effect occurred when "something" collided with the electrons and only ejected them if "something" contained enough energy. He concluded that light consisted of particles, specifically photons. This revolutionary suggestion—discounted at the time by prominent physicist Robert Millikan as "bold, not to say reckless"—earned Einstein the Nobel

> Prize in Physics in 1922. Millikan eventually accepted Einstein's theory and won the Nobel Prize in Physics the following year for his experimental verification of Einstein's ideas.

For imaging devices, the most important aspect of the photoelectric effect is that it illustrates how light can be converted into electrical power. In a photo-conductor, light causes electrons to flow, which creates a current; in a photovoltaic cell, electrons become separated to create a voltage difference; and in a photo-emissive device, like a vacuum tube, electrons are ejected and can be used in imaging devices like photomultiplier tubes. You may think of film-based cameras as old-fashioned, but they use the photoelectric effect just like the CCD does.

The silver bromide crystals in the film that capture light are photo-conductive. Light gives enough energy to the silver bromide to cause an electron to become free from the bromide ions in the salt. This electron travels through the salt grain until it reaches the surface and combines with a silver ion to form metallic silver.

How Not to Make a Digital Camera

You can see intuitively now the essence of a digital camera: Take several million of these chips of photosensitive silicon (the pixels), arrange them in a grid, and capture an image. As always with engineering, the devil lies in the details.

The first attempt to make digital imaging devices, pioneered by RCA, employed the most obvious method: Use wires to connect the pixels in an *x-y* grid. Light striking the pixel caused a charge to accumulate proportional to the light's intensity.

To read that charge, engineers attached to each photosensitive pixel an electronic "gate" that controlled whether the stored charge could flow out of the pixel. By sending signals vertically down the grid and then horizontally across the grid, the charges stored—and thus the image—could be read pixel by pixel. In principle this would work, but in practice this method presents huge problems. All of the tiny electronic components attached to each pixel had a small capacitance; that is, they stored a little charge. So when the signal traveled down the columns, it acquired this tiny bit of charge from each of the other pixels it passed. While the charge added to the signal at each step was small, by the end, the additional charges resulted in significant distortion of the image. This phenomenon, which is known as capacitive coupling, introduced electronic noise that caused striations and patterns to appear in the image. Even worse, this distortion increased as the number of pixels increased. Early x-y photo-grids had 180 columns and rows and produced images with significant noise—imagine using one to replace one of today's CCDs that has 1,600 rows and 2,000 columns! A CCD solves this problem in a simple way: The pixels have no wires attached to them!

Eight Amazing Engineering Stories

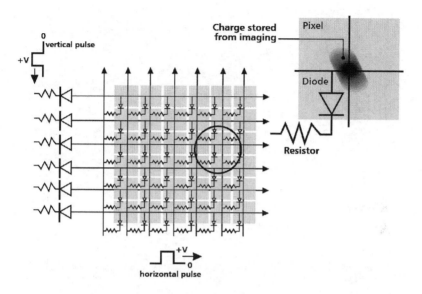

RCA engineers pioneered the use of photoconductive elements to record images. They arranged them in an x-y grid—the light gray areas are pixels, connected by a grid of wires. Charge accumulates at the center of each pixel, where the wires cross at its center. To read the charge on each pixel, pulses in the x and y directions open and close the diodes. As the x (horizontal) pulse moves from left to right, it consecutively opens the diodes. (Diodes allow only current to flow in only one direction; the positive voltage of the pulse opens the diode.) The y pulses work the same way. The horizontal scan rate is much faster than the vertical one. That is, the vertical pulse opens a diode attached to a row, and then the horizontal pulse rapidly zips across that row, opening each pixel in that row. Then the vertical pulse moves down a row, and the horizontal pulse repeats its motion.

Capacitive Coupling

Capacitors in a circuit store energy. Now, the word storage implies that capacitors introduce a time-varying element in a circuit. Current flowing in a circuit with only resistors flows at a constant rate, but introduce a capacitor, and the flow can suddenly start or stop. While many circuits have capacitors built into them, capacitance also shows up unwanted in electronic systems, causing noise that distorts signals. For example, two unconnected wires side by side will become "capacitively coupled" if they are close enough. In digital systems, such as an x-y-readout imaging device, elements become so small that these effects can be large: The mess of wires creates unpredictable crosstalk that just appears as noise. One can see how unwanted capacitance muddies a clear signal by looking at a square wave pulse.

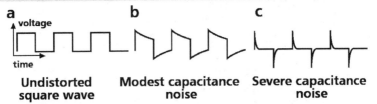

a) Undistorted square wave **b)** Modest capacitance noise **c)** Severe capacitance noise

a) If there is very little or no capacitive coupling, the square wave is undistorted; b) if the resistance times the capacitance in the line equals the period of the square wave, we get significant distortion; c) if the resistance times capacitance far exceeds the wave's period, we get severe distortion, only seeing the spikes where the square wave changes voltage.

CCD: Charge Coupled Device

In a consumer camera, the CCD is about 5 mm long by 4 mm wide and consists of mostly a single slab of silicon. Silicon is a wonderful material for this "monolithic" construction because it can be made insulating, conducting, or semi-conducting by adding other elements to it. To make the pixels within the slab, engineers start by creating insulating sections called channel stops; these divide the slab into pixels in one direction. Next, electrodes that run perpendicular to these channel stops are laid down on the silicon. A pixel, then, is a section of silicon bounded by two channel stops in one direction and three metal strips in

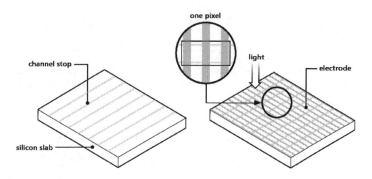

A CCD is created by "doping" silicon with a small amount of boron, which effectively adds a positive charge carrier to the silicon. Engineers then create photosensitive sections within the CCD by adding arsenic to the silicon in between the channel stops. The arsenic adds a negative charge carrier to that region of the silicon. They cover this surface with a thin layer of insulating silicon dioxide, and then engineers deposit thin strips of metal, typically aluminum, perpendicular to the channel stops. Note that a pixel, as shown in the insert, is defined by three electrodes and two channel stops.

the other. In a moment, we'll show why three strips are critical to how the CCD operates.

Just like the silver halides in film cameras or the *x-y* grid in early digital devices, light strikes the surface of the CCD when the picture is taken. It passes through the gaps in the electrodes, allowing the silicon pixels to build up charge. After exposure, the CCD has captured an image—the only problem is that it is stored as charges on the surface of a slab of silicon. To take these groups of stored charges and turn them into an image requires removing the charges from the pixels.

Recall that a camera has two essential functions: to capture and record an image. Right now, we have the image stuck as charges in the pixels. The great innovation of the CCD was how it moved the image from the pixels to the camera's electronics without using external wires and gates that distort the image. To record the image, the CCD shifts the built-up charges from row to row until it reaches the bottom, where a read-out register transfers the charge to the camera's electronics, which in turn construct an image.

In a modern CCD, this process transfers the charge with an amazing 99.9995% efficiency; that is, with very little distortion. If we examine the CCD's edge, looking down the metal electrodes, we can see how the CCD traps and moves charges. The CCD is a metal oxide semiconductor (MOS). It consists of a semiconductor (the silicon) covered first by an insulating layer and then by a thin layer of metallic electrodes. By varying the voltage placed across this device by the electrodes, we can create a trap for electrons, which are the charges that make up the image.

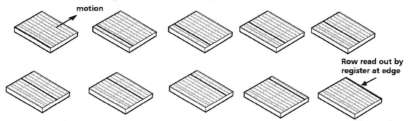

The great technological advantage of the CCD, when it was first introduced, was the way it moved a captured charge. Rather than using wires as in an x-y grid, it instead moves the electrons captured by exposure to light row by row through the solid silicon. (The next two figures describe in detail how this happens.) Highlighted here is a single row (although all rows move) that is transferred down the CCD slab until a read-out register at the bottom records the charges.

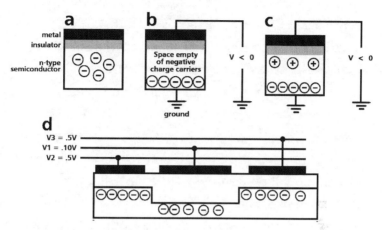

The CCD slab is a MOS (metal oxide semiconductor), which can be used to trap a charge by varying the voltage. To create a pixel, we use three of these MOS structures side by side.

A) If we apply no voltage to the MOS, mobile negative charge carriers are distributed throughout the semiconductor.

B) Applying a negative voltage to the metal moves electrons away from the metal-semiconductor interface.

C) Applying a highly negative voltage drives electrons deep into the bulk of the semiconductor, leaving positive charge carriers near the surface.

D) Using three of these MOS structures side by side allows us to create a "trap" for electrons.

By lowering the voltage of the center electrode relative to the sides, we form a region with positive charge. When the light strikes the silicon, electrons that are trapped in this small potential well flow into the area near the surface. This creates isolated charges on the surface of the CCD's silicon, charges that are located at different points across the CCD's "grid" and make up the image. Now let's turn to recording that image, or the details of getting the charge out of the pixels.

Let's look at a four-pixel section plus one-third of a fifth pixel. As noted above, when light strikes the center wells, each pixel stores a charge. We then drop the voltage on the electrode to the right of each well, allowing the stored charge to migrate to the area with the lowest voltage. We then raise the voltage on the original well, trapping this charge under the next electrode. All charges have now moved to the right by one-third of a pixel. We continue this process until all of the charges reach the bottom edge where the camera's electronics record the information from each row.

Here's what happens in detail. In the figure above, the four pixels are shown at time 1, immediately after exposure. From left to right, the pixels have two charges, three charges, one charge, and four charges trapped at the semiconductor-metal interface. To move the charge one row to the right, the following happens: At time 2, the potential in each well immediately to the right of the stored charges drops to the same voltage as in the well to its left. This causes the charges to move to the right. Then at time 3, the potential of the well that originally trapped the charges rises (Notice that a charge on the left is moving from a well not seen in time 1). At time 4, the potential in each well immediately to the right of the stored charges drops to the same voltage as in the well to its left. The charges once more move to the right. Then at time 5, the potential of the well that previously trapped the charges rises. At time 6, the potential in each well immediately to the right of the charges that were just moved and stored drops to the same voltage as in the well to its left. The charges once more

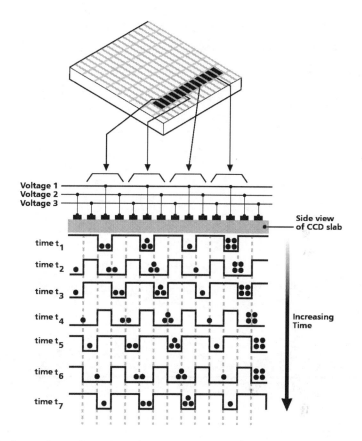

This figure shows four pixels from four different rows of a CCD and what happens at seven sequential times to these four pixels.

move to the right. Then at time 7, the potential of the well that previously trapped the charges rises. This completes one "clock" cycle: The charges in the rows have moved down a row. This continues until all of the charge is removed from the CCD.

It may seem like a cumbersome process. Indeed it can be very slow because it is serial; that is, there is no skipping or jumping by rows. The first row must be transferred first, then the second, and the third, and so on. You can observe this row-by-row serial reading of an image in the time you need to wait between taking pictures with a digital camera. This time delay from the serial motion is the price paid to have no wires as in an *x-y* device, where individual pixels can be read. The great virtue of the CCD is that it captures a clean, clear image.

How a CCD Creates Color

A CCD only detects light intensity, not color. This means a single pixel would measure the combined intensity of all light colors at the same time. We could use this to make a black-and-white image, but to make a color image we need to separate the entering light into its three components: the primary colors red, green, and blue. It would then seem obvious to record a color image using three CCDs.

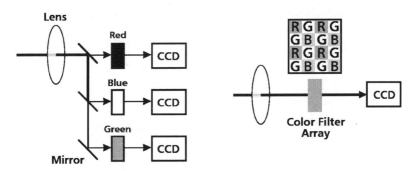

Left *Engineers could use three CCDs to create a color image, but this would be very expensive, so instead they use* Right *a single CCD covered with a color filter array. This creates an image that is a mosaic of red, green, and blue sections. Full color is restored using an algorithm. Shown here is the Bayer Color Filter Array.*

Using a prism, we could split the incoming light into three rays and pass one through a red filter, one through a green filter, and the last through a blue filter. A CCD behind each would record the intensity of each color, and then we could recombine them into a color image. Although this solution is simple, engineers rarely use such a design in consumer digital cameras because it's far too expensive and requires three CCDs! Instead, they use a bit of math to create the color image. A typical consumer camera uses a single CCD with a color filter array (CFA) placed in the path of the incoming light. Engineers divide this array into pixel-sized sections: Some of the sections have a red filter, some green, and some blue. This means that the image that comes out of the CCD is a mosaic of red, green, and blue sections. The camera then applies an algorithm to estimate the correct colors and fill in the picture. For example, if a red filter covers a pixel, we would need to estimate the green and blue components of that pixel. To do this we'd use the adjacent pixels. For instance, if the intensity of green color in a nearby pixel is 5% and the intensity of green in another nearby pixel is 7%, a good estimate of the green intensity in the pixel of interest—the one covered by the red filter—would be 6%. This process works because the image's significant details are much larger than the pixels. Sounds impossible, but you've seen the results yourself: Every digital picture you've ever snapped likely uses this method!

Active Pixel Sensors (APS)

Your smartphone likely doesn't use a CCD; it's probably an APS (active pixel sensor) CMOS chip. In spite of the CCD's great advantage over the first x-y pixel devices, the CCD still has disadvantages. This should be no surprise because every engineered object is a balance or trade-off among desired properties. The quest to make both larger imaging devices (telescopes, for example) and smaller ones (cell phones) highlights the downside of CCD technology. Making large CCDs calls for more efficient row-to-row charge transfer. Every time a transfer occurs, a little bit of charge is lost. In a 1024 x 1024 CCD, for instance, about 1% of the charge is lost from the last pixel read. (The last pixel is the worse-case scenario: It must be transferred 1,024 times to the readout register, then down that register another 1,024 steps. For these 2,048 steps, the amount of charge transferred would be $0.9999995^{2048} = 0.989$ charge transferred or about 1.2% lost). In an array that is 8192 x 8192, this same efficiency would mean about 8% would be lost from the farthest pixel. So, to make the 8192 x 8192 CCD have the same amount lost as the 1024 x 1024 array would require somehow increasing the charge transfer to 99.99993%. One can see that eventually the size of the CCD will outstrip any attempts to make the efficiency higher. In addition, the CCD is slow to read because it is discharged row by row. In making tiny cameras for cell phones, the CCD has two main limitations. First, the CCD has to integrate onto a chip with other components. Second, the CCD requires large voltages, perhaps 10 to 15 volts, that can drain a cell phone battery. Oddly, the way forward for imaging devices is to, in a way, return to the *x-y* devices that lost out to the CCD in the 1970s.

The early devices suffered from severe capacitive coupling that distorted the images recorded. These early devices were passive pixel sensors (PPS). New chip-making methods allow production of *x-y* devices with a transistor built into every pixel. In this APS device, the transistor functions as an amplifier that increases the signal from a pixel, thus overcoming the noise from capacitance. In addition, the transistor allows digital filtering techniques to be used to reduce noise, something that could not be done in the early PPS system.

> ### The CCD and the Nobel Prize
>
> In 2009, the Nobel Committee awarded the Nobel Prize in physics to Willard S. Boyle and George E. Smith for "the invention of an imaging semiconductor circuit—the CCD sensor." In a way, their work on the CCD was inspired by magnetic bubble memories, which was all the rage in 1969 at Bell Laboratories. These memories used tiny magnetized spots, called bubbles, moved about by currents to store information. This worried Boyle. As executive director for the division of Bell Labs that worked on silicon—the dominant medium for computer memory at the time—he worried that these new bubbles might divert funding and support from silicon research. Boyle invited his friend and colleague George Smith to help him come up with a competitor to this new technology. On a chalkboard, they devised a way to use silicon, silicon dioxide, and metal electrodes to store charge in specific areas on the surface of the silicon. The conversation took about an hour as they jotted down in their notebooks that the device could be used as "an imaging device" and a "display device" in addition to being used for computer memory. After reflecting on their chalkboard talk for a few weeks, they decided to build a prototype CCD. Within a week, they had a working device that proved the concept of their idea. They did not pursue using the CCD to capture images; that task fell to their Bell Labs colleague Michael Tompsett.
>
> A great engineer, Tompsett carefully turned the brainstorm of a charge-coupled device into a reality. His name alone appears on the first patent for the CCD as an imaging device; the patent was titled, appropriately enough, "Charge transfer image devices." While one

might feel Tompsett was overlooked for the Nobel Prize, the prize is generally given for the invention or discovery of fundamental concepts in physics. Boyle and Smith indeed laid down the idea of a CCD, but controversy arises because the only practical application of a CCD is as an imaging device.

How a Smartphone Knows Up from Down

Humans and other living things can tell up from down because we feel the pull of gravity. For instance, if you drop this book, gravity will accelerate it toward the center of the earth until it comes to rest on the ground. But how does a smartphone know which way the phone is rotated, so the screen is always right-side up? To ensure that the screen on your smartphone always is pointed in the correct direction, engineers place tiny accelerometers inside the phone to orient it with respect to the earth. An accelerometer is a device that measures gravitational pull. But how can these tiny accelerometers tell which way is up?

Basics of an Accelerometer

You can see the essential principles of an accelerometer in the simple device shown in the next figure. When that device is upright, gravity stretches the spring downward, as indicated by the mark labeled 1g, meaning one unit of gravitational acceleration. (One g is what you feel with little or no other acceleration on earth; for reference, when a roller coaster car takes off you feel about 2.5gs, and fighter pilots black out at about 10gs.) On the device 0g occurs when the tube lays flat so the spring feels no gravitational pull, and the spring has no extension. This distance from 0g to 1g sets the scale for marking 2g, 3g, etc.

We can use this accelerometer to measure the upward vertical acceleration experienced by the tube. If we were to quickly jolt the tube upwards, we would see the weight drop inside the tube, possibly to 2 or 3gs. Three of these basic accelerometers can be used tell us the orientation of an object.

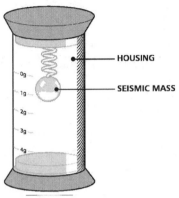

A simple accelerometer: A glass tube with cork stoppers at each end makes up the housing. The seismic mass is a lead ball tethered to the housing by a spring.

In an oblong box, align one accelerometer each along the *x*, *y* and *z* axes. By measuring changes in the lengths of the springs, we can detect which edge points up relative to gravity. In the first position, the *x*-axis accelerometer records 1g, while in the *y* and *z* direction the weights lie against the side of the tubes and the springs are not extended. Rotate the box so it sits on its long edge, and the *y*-axis accelerometer will register 1g, while the *x* and *z* accelerometers will read 0g. Although the accelerometer inside a smartphone is a bit more complex, it works using these same principles.

Eight Amazing Engineering Stories

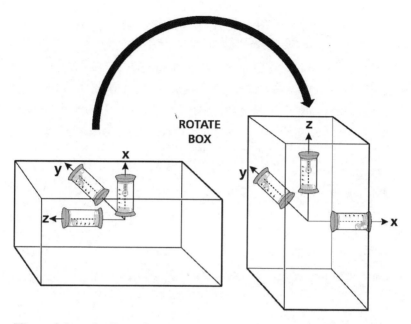

Three of these simple accelerometers can determine the orientation of a box. Note the change in the x and z accelerometers. When the box lies on its z-axis, the ball in the accelerometer along that axis lies flat while the spring in the accelerometer on the x-axis is extended. When the box is rotated, the ball in the x-axis accelerometer lies flat, while the ball stretches the spring in the accelerometer along the z-axis.

Technology & Privacy

We marvel at the great things technology can do, like detect the subtle motions of a cell phone. As with all technologies, though, power comes with peril—even with such a seemingly innocuous object as a smartphone accelerometer. Researchers at the Georgia Institute of Technology showed that such an accelerometer could track the keystrokes of a computer user. They placed a phone on the table beside a computer keyboard. Since the accelerometer samples a phone's vibration about 100 times per second, it would jiggle slightly with each press of the neighboring keyboard. (The keyboard moved the table slightly, and the ultra-sensitive accelerometer detected these small vibrations of the table.) By detecting pairs of keystrokes located relative to each other on the keyboard, they could guess with 80% accuracy the word being typed. For example, in typing "canoe" the user would create four keystroke pairs: C-A, A-N, N-O, and O-E. The accelerometer would determine for those pairs if the two letters were related in one of four ways: Left-Left-Near, Left-Right-Far, Right-Right-Far, and Right-Left-Far. They compared this to a preloaded dictionary of likely words to determine the most probable word made for these pairs of keystrokes. Small wonder, then, that Melvin Kranzberg, a historian, once observed "technology is neither good nor bad; nor is it neutral."

The Accelerometer Inside a Smartphone

The figure below shows a typical smartphone accelerometer. It's very small, only about 500 microns long on each side. The housing, which is stationary, is the large block at the base, to which are attached several stationary polysilicon fingers. The seismic mass is the roughly H-shaped object with the "tongues" extending from it; it's tethered at the ends so it can jiggle left and right between the stationary fingers.

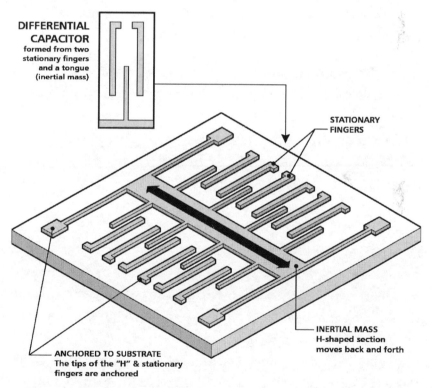

A typical smartphone accelerometer made from a slab of silicon.

Recall that in our simple accelerometer with the weight and spring arrangement, we measured the acceleration of the box by how much the weight moved relative to the tube. This device measures acceleration by the degree to which one of the "tongues" hanging off the H-shaped section moves relative to the two stationary fingers. We did this by eye for the weight and tube accelerometer, but here we use the electronic properties of silicon. The tongue and two stationary fingers form a differential capacitor, a device that stores charge. As the accelerometer moves, the charge stored within the differential capacitor changes, causing a flow of current. Through careful calibration, engineers can link the magnitude of the current flow with the pull that the accelerometer feels from gravity.

Using a Capacitor to Make an Accelerometer

We can make the simplest capacitor from two metal plates with an air gap between them, as shown in the next figure.

If we hook these capacitors to a battery, current will flow as charge builds up on the plates: positive on the top plate and negative on the bottom. Once enough charge builds up, the current stops because the battery isn't powerful enough to force charge across the gap. For most well-made capacitors, this situation will persist until we change something in the circuit. For example, if we move the plates a little closer, then current will flow again as the charge is redistributed. You can picture how a capacitor would be useful in an accelerometer: Imagine one plate as the housing and the other as the seismic mass. When we hold that accelerometer stationary, no current flows. But move the seismic mass, or the top plate, and a current now starts to flow. Just as we did with the weight and tube accelerometer, we can calibrate this: Subject the two-plate capacitor accelerometer to known accelerations and measure the currents that flow.

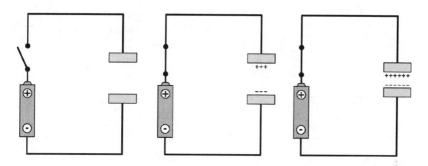

The leftmost image shows a two-plate capacitor before the circuit is closed. There are no charges on the two metal plates. When the circuit is closed (center image) a charge builds up on the plates and current stops flowing. If the plates are moved closer together, as in the image on the left, current flows until the charge is redistributed on the plates.

A smartphone accelerator doesn't quite work like this, but we're getting close. The problem is that our two-plate capacitor has a serious defect.

In the simple two-plate capacitor, the relationship between plate position and current is non-linear. For example, if we reduce the distance by 25%, the capacitance will drop by 6%. If we again change the distance by 25%, the capacitance drops by about 4.5%. This makes the accelerometer difficult to calibrate. Engineers prefer a linear response; that is, if we reduce the distance by one-quarter and then by another quarter, we'd like to see the same drop in capacitance each time. This would allow for a more uniform sensitivity across the useful range. So if we set up a slightly different capacitor where the top and bottom plates are stationary and the middle one moves, the current generated by changes in this differential capacitor will be linear.

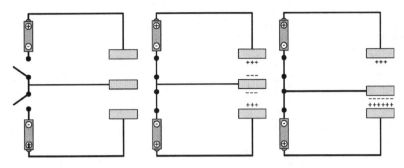

In this differential capacitor, we measure the difference in the charges that form in the bottom capacitor (the bottom and middle plate) and the top capacitor (the middle and top plate). On the left, the capacitor is uncharged. If we use the same size batteries and keep the middle plate exactly between the top and bottom plate, then positive charge (as shown in the middle image) will build up equally on both plates. If we move the middle plate closer to the bottom plate, the capacitance of the bottom capacitor increases while the top decreases. The difference in capacitance is linear with respect to the motion of the middle plate.

Look at what happens if we charge the differential capacitor using two batteries with opposite polarities: Current flows until the middle plate becomes negative and the top and bottom plates are positive. Now, if we move the middle plate, current will flow to redistribute the charge. If we measure that flow only between the middle and bottom plate, it will be linear with the middle plate's motion. (At the end of the chapter, there is an explanation of the mathematics of capacitors.) This is exactly how the accelerometer works: The tongue from the H-shaped piece corresponds to the middle plate, and the two stationary fingers are the fixed plates of the differential capacitor. At rest, no current flows. But if we move the accelerometer, the tongue will jiggle, creating a current proportional to the acceleration. Now we have the perfect device to put in our phone and tell us which way is up, but that is only half the battle. The next problem lies in making something so small but so complex.

How to Make an Accelerometer

It would seem nearly impossible to make such an intricate device as the tiny smartphone accelerometer; at only 500 microns across, no tiny mechanical tools could craft such a thing. Instead, engineers use some unique chemical properties of silicon to etch the accelerometer's fingers and H-shaped section. The method is known as MEMS, or micro electro-mechanical systems. To get an idea of how they do this, let me show you how to make a single cantilevered beam out of a solid chunk of silicon. A cantilevered beam is one that is anchored at one end and sticking out over a hole; for instance, a diving board is a cantilevered beam. The moving section of the smartphone accelerometer is just a complex cantilevered beam. Let's start by etching a hole in a piece of silicon.

Empirically, engineers noticed that if they poured potassium hydroxide (KOH) on a particular surface of crystalline silicon, it would eat away at the silicon until it formed a pyramid-shaped hole.

To make a pyramidal hole in silicon, engineers first cover all but a small square of the (100) plane—that's what we meant by a "particular surface"—with a "mask" of silicon nitride (Si_3N_4). The mask is impervious to the KOH, so the KOH etchant will now only etch within the square shape cordoned off by the mask until it's washed away.

This directional etching occurs because of the unique crystal structure of silicon.

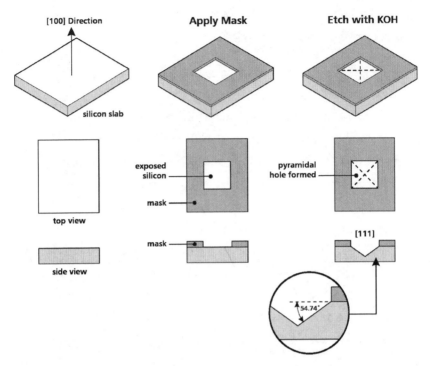

The three main steps (from left to right) to make a pyramidal hole in a piece of crystalline silicon. Each step is shown from three angles (top to bottom): perspective view, top view, and side view of each step.

In a chunk of silicon, each silicon atom has four closest silicon neighbors surrounding it. The four neighbors create a tetrahedral region around the silicon at the center. Millions of these "units" make up a slab of silicon. They're packed together in an open structure that is not the same in every direction. For example, the number of atoms along the three directions shown above—labeled [100], [010], and [001]—differ in the number of atoms in the planes perpendicular to each direction. The atoms are packed more tightly along the [111] direction than in the [100] direction. This means that the KOH dissolves both the (110) surface and the (111) surface, but it chews through the latter much faster. This is why it makes a pyramidal hole.

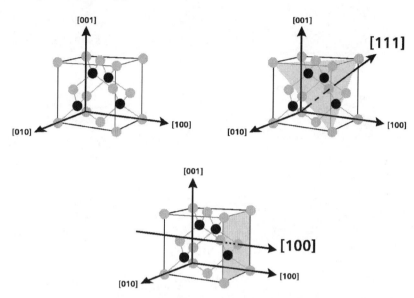

Crystal structure of silicon. Every atom in a silicon crystal is identical. Each is bonded to four other silicon atoms. The darkened atoms in this figure show most clearly that each atom has four nearest neighbors. Note that there are more atoms in the plane perpendicular to the [111] direction than there are in planes perpendicular to the [100] direction.

To make a cantilevered beam, engineers mask all the surface except for a U-shaped section. At first the KOH will cut two inverse pyramids side by side.

As the etching continues, the KOH begins to dissolve the silicon between these holes. If we wash KOH away at just the right point before it dissolves the silicon underneath the mask, it will leave a small cantilever beam hanging over a hole with a square bottom.

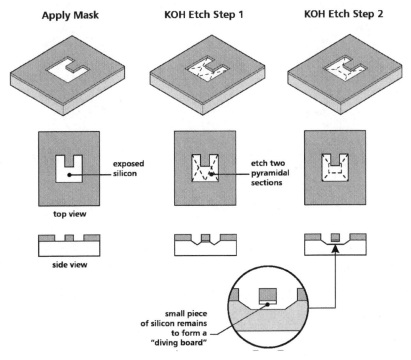

The three main steps (left to right) to make a cantilevered beam in a piece of crystalline silicon. Each step is shown from three angles (top to bottom): perspective view, top view, and side view.

Although engineers use these principles to produce smartphone accelerometers, they use much more complicated masks and multiple steps. As you can picture, first a machine would apply a mask, then etch the silicon, change the mask, and etch some more; the process is repeated until they create an intricate structure. While complex, a key point is that the whole process can be automated. Engineers now make all sorts of amazing things at this tiny scale: micro-engines with gears that rotate 300,000 times a minute; nozzles in ink-jet printers, and, my favorite, micro-mirrors that redirect light in optical fibers.

Eight Amazing Engineering Stories

In Depth: The Mathematics of Capacitors

Simple algebra reveals why the capacitance of a two-plate capacitor changes non-linearly with a change in the plate separation and why a differential capacitor changes linearly.

The capacitance between two plates, which is just the amount of charge that it can hold, depends on three factors:

1. the distance between the plates,
2. the surface area of the plates,
3. the electrical properties of the material between the plates, known as the dielectric.

The dielectric constant (ε) is the ratio between the amount of electrical energy stored in a material by an applied voltage to the energy stored if the space between the plates were filled with a vacuum. Often this space is instead filled with a common mineral known as mica, which has a dielectric constant of around 5; frequently the dielectric between the plates is just air ($\varepsilon = 1$). We define capacitance as

$$C = \frac{\varepsilon A}{d}$$

Where ε is the dielectric constant, A is the area of the plates, and d is the distance between the plates.

If we move the plates closer together by a small amount, say Δ, then the new capacitance is

$$C = \frac{\varepsilon A}{d - \Delta} = \frac{\varepsilon A}{d(1 - \frac{\Delta}{d})}$$

For very small motions, we can use a Taylor Series to approximate this:

$$C = \frac{\varepsilon A}{d}\left[1 + \frac{\Delta}{d} + \left(\frac{\Delta}{d}\right)^2\right]$$

The Taylor Series is a useful way to simplify some mathematical formulas under certain conditions. For example,

$$\frac{1}{(x+y)}$$

can be approximated by

$$1 - \frac{y}{x} + \left(\frac{y}{x}\right)^2$$

if y is much smaller than x. You can test for yourself how much smaller y must be than x for this to be true. Here, we are assuming that the change in the distance between the two plates will be much smaller than the original distance, so we approximate with a Taylor Series.

That squared term causes a non-linear response, which is most easily seen in the plot below.

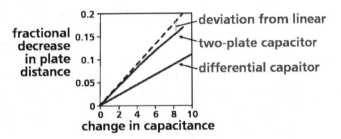

This plot shows the response of both a two-plate and a differential capacitor. The y-axis is the fractional change in the separation of the plates—here they are getting closer together—and the x-axis shows the change in capacitance. Note how the two-plate capacitor deviates from a linear response, while the differential capacitor has a linear response in the same range.

To overcome the non-linear problem, engineers use a differential capacitor. As described in the chapter on the accelerometer, this capacitor has three plates: one in the center, one above a distance d, and one below a distance d. The capacitance of each plate is then:

$$C_{top} = \frac{\epsilon A}{d}$$

and

$$C_{bottom} = \frac{\epsilon A}{d}$$

Of course if the plate is perfectly centered, the difference between these two capacitances would be zero, but it will change if we move the middle plate down by Δ. That is, as the center plate gets further from the top plate by Δ and closer to the bottom plate by Δ. This difference becomes:

$$C_{bottom} - C_{top} = \frac{\epsilon A}{d-\Delta} - \frac{\epsilon A}{d+\Delta}$$

Using that Taylor Series expansion for each capacitance, we get:

$$C_{bottom} - C_{top} = \frac{\epsilon A}{d}[1 + \frac{\Delta}{d} + (\frac{\Delta}{d})^2 - 1 + \frac{\Delta}{d} - (\frac{\Delta}{d})^2]$$

The two squared terms cancel each other out, leaving us with:

$$C_{bottom} - C_{top} = \frac{(\epsilon A)}{d}\frac{2\Delta}{d}$$

As you can see in the plot above, the capacitance now changes linearly with plate separation.

Cesium
[seez-i-uhm] | Symbol: Cs
Atomic number: 55 | Atomic weight: 132.9054 amu

In 1861, Robert Bunsen was on a search for a new element. Having already invented his famous burner, Bunsen was working with co-discoverer Gustav Kirchhoff (famous today for his eponymous laws of electricity) to extract a new element from mineral water. Boiling down 30,000 liters of water, they extracted from it everything they already knew about—lithium, sodium, potassium, magnesium, and strontium. What remained was a thick "liquor" of some unknown substance, which Bunsen placed in the flame of his burner. Two brilliant blue lines appeared in the flame when viewed through a prism, indicating a new element. He named it cesium, from the Latin caesius, *meaning blue-gray. The defining characteristic of this soft, silvery metal is its low melting point of 29 ºC. Place some in your hand and it'll melt, but it is so reactive that it will also* burn your palm: The moisture in your skin will be decompose into hydrogen and oxygen.

Cesium has two traits important in making the first atomic clock, as described in this chapter. First, its low melting point makes it easy to create a gas. Second, the element has only one stable isotope (^{133}Cs), so all cesium atoms are identical, allowing it to be used as a standard in timekeeping.

How an Atomic Clock Works

INSIDE A BUNKER IN Boulder, Colorado, an atomic clock—blandly named NIST-FI—measures time so accurately that it will neither gain nor lose a second over the next 100 million years. It sends its time signals to an elegant eighteenth century building, the Pavillon de Breteuil on the outskirts of Paris, which houses the *Bureau International des Poids et Mesures,* (*International Bureau of Weights and Measures*) keeper of the SI (*Système International d'Unités*). There, an international team of clock watchers combine the results of almost 260 clocks to create "Coordinated Universal Time" (UTC), the world standard for civil timekeeping.

Our high-tech world depends on accurate time in nearly every respect, but perhaps the most striking example is the global positioning system (GPS), which requires time be measured to an accuracy of at least one nanosecond to operate correctly. To achieve that accuracy, every GPS satellite orbiting the earth contains an atomic clock that marks time with incredible accuracy.

To make a clock extraordinarily accurate requires a standard that will remain the same over long periods of time. For the atomic clock, it is easy to specify that standard: The energy difference between the ground states of a cesium atom—a difference of 3.8 electron volts, or about six-sextillionths of the energy used when a 100-watt incandescent bulb burns for one second. This energy difference is the same for every atom of cesium kept at the same conditions of pressure and temperature. The hard part is engineering a device that uses this standard to make a reliable, useful clock. To understand the engineering principles used in designing an atomic clock, we need first to understand how to keep time by resonance.

Basics of Modern Timekeeping

In the past, time was measured using physical actions, such as the flow of water in elaborate sixth century BC Chinese clocks, or the hourglass's flowing sand. In our modern age, we measure time by resonant control. (See figure for a definition of resonance.)

The most familiar example of using resonance for timekeeping is the pendulum of a grandfather clock: It swings back and forth with a regular period, typically two seconds for the pendulum to complete one arc. (The period is the time it takes for the pendulum to swing completely in one direction and then back again to its original starting point.) Based on this period, the manufacturer designs the gears in the clock to translate this periodic motion, or resonance, into a movement of the hands that marks time in minutes. If there were no friction or drag, the pendulum would swing forever, keeping perfect time; but, of course, the arc slowly decays, that is, it gradually stops swinging. On each swing, the pendulum loses a little energy to friction and does not return to as high a point as it did before. If you wanted a grandfather clock to keep better time, you could tap it every time the pendulum's arc decreased just a little bit to put it back on track.

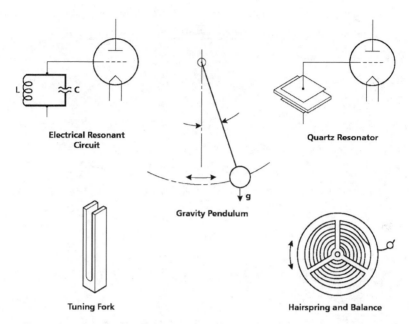

Resonance is the tendency of a system to oscillate with larger amplitudes at some frequencies than at others; these points are called a system's resonant frequencies. All resonators, if moved from rest, transform their stored energy back and forth from potential to kinetic at a rate depending on the mass and stiffness of the spring or pendulum, or equivalent electrical properties. At each oscillation, that is, the change from potential to kinetic energy, the resonators lose a small portion of their energy to internal friction and so eventually decay. There are many ways to create resonance, all of which can be used to tell time. Shown in this figure are a) an electric circuit with an inductor and capacitor; b) a pendulum driven by gravity; c) a quartz resonator; d) a tuning fork made of steel or quartz; and e) a hairspring with a balance as used in early twentieth century watches.

In essence, this is precisely how an atomic clock works. It also has a resonator keeping fairly good, although not perfect, time. Taking the place of the pendulum inside the clock is a piece of quartz, and the "tap" comes from an electronic signal guided by a device that uses cesium atoms to detect when the resonator's period has decayed a tiny bit. Picture the "atomic" part like cruise control. In a car, you use cruise control to set a constant speed, and if the car speeds up or slows down on a hill, the control system responds automatically to keep the car's speed at its set point. In an atomic clock, the speed is the period of the resonator. If it slows down, a circuit tells it to speed up a bit and vice versa. In engineering parlance, this is a type of feedback control.

As you might guess, creating the "cruise control" from a cesium atom is pretty complex. Let's break it down by looking at the resonator that lies at the heart of an atomic clock.

Quartz Resonators

A pendulum is only one of many ways to create resonance for measuring time. A pendulum system would be far too inaccurate for an atomic clock, so engineers use a piece of quartz, typically just a solid, rectangular slab a few millimeters in length and width and fractions of a millimeter in thickness.

It doesn't seem obvious that it would work as an oscillator, at least not as obvious as a pendulum, yet the principles are the same. For a moment, instead of thinking about a slab of quartz, think of a rectangular chunk of Jell-O. If you tap it, it vibrates back and forth. It doesn't last for long, but it does have a periodic motion or resonance, and just like the pendulum, its motion stops eventually. A piece of quartz will resonate in the same way.

Eight Amazing Engineering Stories

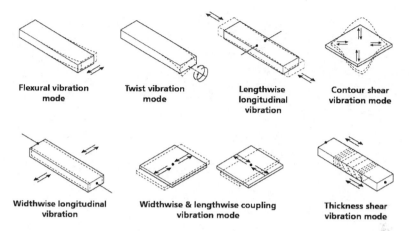

A piece of quartz can oscillate (vibrate) in many different ways, each of which has a different frequency depending on the thickness and dimensions of the crystal. Shown above are some of its main modes of vibration. Most important for the atomic clock is the thickness shear vibration mode (bottom right in the figure). The most precise crystals vibrate in the fifth overtone of this mode at frequencies of 5 MHz or 2.5 MHz.

Tap it in the right way (see next figure), and it will oscillate at five million times per second, or 5 MHz. Unlike Jell-O, we don't tap the quartz to have it oscillate. To start the oscillation of the quartz to measure its period, engineers use the piezoelectric properties of quartz. The word *piezoelectricity* comes from the Greek verb *piezein*, which means to squeeze or press.

To start the quartz crystal oscillating, we "tap" it with a jolt of electricity (a small voltage across the quartz) and then measure the vibration from the current it produces. That tap will keep the quartz resonator going for a long time; for example, its oscillations will last roughly 1,000 times longer than the swing of a pendulum of a grandfather clock.

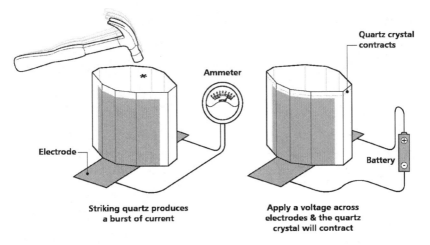

Striking quartz produces a burst of current

Apply a voltage across electrodes & the quartz crystal will contract

A piezoelectric material can convert mechanical motion into electricity and vice versa. For example, if we were to attach two foil electrodes to a piece of quartz and then strike it with a hammer, the crystal would generate a current. And if, in turn, we apply a voltage across those electrodes, the crystal will deform.

It might seem the quartz oscillator solves the problem of creating precise time. Quartz is ideal for clocks because of its outstanding physical hardness as well as its mechanical and chemical stability. A quartz resonator provides precision only to about 1 second in 300 years for a short period of time (although the best resonator can achieve 1 second in 3,000 years), but if we combine it with feedback from an atomic standard, we can make a clock with astounding accuracy.

Cesium-based Atomic Clocks

Recall that our idea was to take an oscillator and, just as the period of its motion begins to decay, give it a "tap" to restore its oscillations. The "atomic" part of an atomic clock uses cesium to create a way to determine when the oscillations of a quartz crystal have decayed too much and the crystal needs to be "tapped." We do this by measuring a particular property of cesium.

The atoms in pure cesium exist mostly in two slightly different forms: A low energy form and one of just slightly higher energy. For an atomic clock, these two variations, usually called states, have two properties critical in making a clock:

1. They can be separated by a magnet because they differ in their magnetic properties.

2. The lower energy atoms can be converted to the higher energy ones if we bombard cesium with radiation of exactly the right value as characterized by the energy's wavelength.

In that word "exactly" lies the heart of the atomic clock's great accuracy. Engineers tie the frequency of the quartz resonator to the wavelength of the radiation bombarding the cesium. That is, if the resonator's frequency changes, then the radiation changes wavelength and no longer converts the lower energy to the higher energy atoms. This means that if we can detect whether higher energy cesium ions are being converted from the lower energy ones then we have a feedback mechanism for keeping the quartz resonator's frequency accurate. Here's how it's done.

In an oven, we heat cesium chloride to create a gaseous stream of cesium ions. The stream contains both lower- and higher-energy-state ions, so we first pass it through a magnet to separate the two types of ions. The higher-energy ions are discarded, and the lower-energy ions are allowed to pass into a chamber. Inside that chamber, we bombard the ions with energy equal to the energy difference between the two lowest ground states of cesium, which converts the lower- to the higher-energy ions. As these gaseous ions leave the chamber, they pass through another magnet that directs high-energy ions toward a detector, discarding any lower energy ones. The detector converts the arriving ions into a current. The key to understanding how this creates an extraordinarily accurate clock is seeing what happens if the energy isn't tuned to just the right wavelength to make the atoms change.

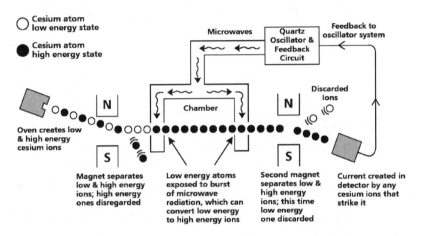

A schematic of the first atomic clock made in 1955. An oven creates a gaseous stream of cesium. This gas contains a mixture of cesium in its lowest and next-to-lowest energy states. The stream flows through a set of magnets that remove the high-energy atoms. The lower-energy atoms flow into a chamber where they are bombarded by microwave radiation whose frequency is tied to the frequency of the quartz oscillator. If that radiation is of the right frequency (or, equivalently, wavelength), all of the cesium atoms in the lower energy state are converted to higher energy. As they exit the chamber, a second set of magnets directs these higher-energy atoms to a detector. The signal from this detector provides feedback to the quartz oscillator: Should the frequency of the oscillator drift, the signal from the detector decreases, indicating the quartz oscillator needs to readjust its frequency (see the next figure for the signal from the detector). In this way, the quartz oscillator can keep time such that it loses only a second or so in a million years.

The trick here is to tie the current from the detector to the quartz oscillator. When the oscillation's period decays, the energy bombarding the cesium ions in the chamber changes, and fewer high-energy ions exit the chamber, so the current decreases or stops.

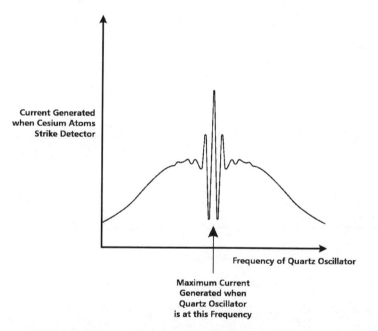

A detector at the end of the chamber measures the number of high-energy cesium ions produced. It converts the number into a current. If we plot that current versus the oscillator frequency, we can see a clear current spike at the optimal frequency. When the electronics of the atomic clock detect this current decreasing, they send a voltage to the quartz resonator and, via the piezoelectric effect, restore its oscillation frequency.

The decrease in current tells the electronics to "tap" the oscillator and correct the period of oscillation. The electronics create this "tap," of course, by applying the proper voltage that, via the piezoelectric effect, deforms the quartz and restores its oscillation frequency.

Uses of Accurate Time

It's very nice to have clocks that keep such extraordinary time, but of what use are they to engineers? Well, an incredibly important use of atomic clocks is our global positioning system, used in almost every facet of life today. The obvious uses of GPS include guiding commercial planes, helping us find a location, and keeping track of a fleet of cargo trucks. However, GPS reaches even deeper into our world, striking at the core of our society: food production. Farmers use GPS extensively for precision or site-specific farming. This technology allows farmers to apply pesticides, herbicides, and fertilizers precisely, which reduces expenses and increases yield. Before GPS, a 4,000-acre farm needed eight or nine tractors; today the same farm needs just three or four.

Throttling a GPS Device

All technological advances come with promise and peril. A GPS navigation unit makes life easier, whether it's a trip for pleasure or tracking a trucking fleet so its operations can be optimized. But in the great, even astonishing, precision of GPS lies the potential to guide a weapon exactly to its target. In the United States, the Department of Commerce requires all exportable GPS devices to shut down when they travel faster than 1,000 knots or when their altitude is above 18,000 meters. This restriction prevents GPS devices from being used to guide missiles to targets.

Eight Amazing Engineering Stories

How the GPS System Works

The global positioning system consists of twenty-four active satellites orbiting the earth, with more added every year (Some satellites are in reserve in case of outages). Maintained by the United States Air Force, they circulate around the earth in about 12 hours at what's called "medium earth orbit;" that is, 20,200 km above the earth. In comparison, geostationary satellites like Meteosat meterology satellites orbit at about twice the distance of the GPS satellites. In some ways, it would be easier to operate the system if the GPS satellites were geostationary, but they would then be too far away to transmit a strong enough signal. A typical GPS satellite contains four atomic clocks: two based on cesium and two on rubidium. Ground stations (called the "Control Segment" of the GPS system) in Hawaii, Ascension Island, Diego Garcia, Kwajalein, and Colorado Springs monitor the satellites' operational health and also transmit to the satellites their exact position in space. For example, the ground station sends corrections to a satellite's position information and also transmits clock offsets to correct any changes in the satellites' atomic clocks. The satellites are synchronized with each other and to a master clock—called GPS time—but only to within about a millisecond. The signals are traveling at the speed of light, so this millisecond means an error of about a million feet or 300 kilometers. The ground stations could update the clocks, but practice has shown it is better to leave the clocks alone and send to the satellite how much its time is offset. These offsets are then sent in the signal from the satellite to the receiver.

Each satellite sends out a signal every millisecond. This signal contains the exact position of the satellite and the time (including correction) that the signal left the satellite. The position information is the satellite's ephemeris, from the Greek word *ephemeris* meaning *diary* or *journal*. On the ground, a GPS receiver detects these signals, recording the time they arrive. Using the difference between the arrival time and the departure time embedded in the signal, the receiver calculates its distance from a satellite: The time difference multiplied by the speed of light (the radio signals travel this fast, attenuated a bit by the ionosphere and troposphere, which we'll ignore here) gives the distance between the receiver and the satellite. In principle, if the receiver knows its distance from three precisely located satellites, it can determine its own position, but as we'll see, it really takes four satellites.

Three satellites would suffice for location, except that the clock on the receiver isn't accurate. We could put an atomic clock in the receiver, but then it would be very expensive and likely quite large. So, to correct the receiver's clock, we use a fourth satellite.

The Mathematics of GPS

When a receiver locates its position based on its distance from satellites of known location, it is solving four equations simultaneously.

To calculate the distance from the receiver to a satellite we use this formula from solid geometry:

$$d_1 = \sqrt{(x_r - x_1)^2 + (y_r - y_1)^2 + (z_r - z_1)^2}$$
$$d_2 = \sqrt{(x_r - x_2)^2 + (y_r - y_2)^2 + (z_r - z_2)^2}$$
$$d_3 = \sqrt{(x_r - x_3)^2 + (y_r - y_3)^2 + (z_r - z_3)^2}$$

Where d_1, d_2 and d_3 are the distances between the receiver and each satellite, which are obtained using the speed of light as

Eight Amazing Engineering Stories

described above; and x_i, y_i, and z_i are the coordinates of the satellites, which are embedded in their signals. Because all three satellites have synchronous time from their atomic clocks, we only have three unknowns: the positions of the receivers x_r, y_r, and z_r. Note that if the satellites weren't synchronized we'd have

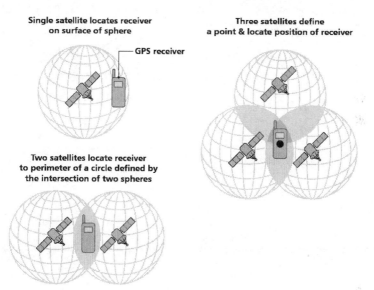

In principle, the receiver can use simultaneous signals from three satellites to determine its position, but as noted in the text it needs a fourth satellite to correct the receiver's clock. Using information from the first satellite's signal —the time it was sent and its position—the receiver can determine that it lies on the surface of a sphere of radius d_1. From the second satellite, it creates another sphere of radius d_2—the distance the receiver is from the satellite. The receiver must be where those two spheres intersect, as shown by the circle represented by the shaded area. The location must lie on that circle. The distance from the third satellite creates another sphere, whose interaction with the first two spheres narrows the receiver location to two points: One can be rejected immediately because it is far from the earth's surface, the other is the correct location of the receiver.

three more variables, an insolvable problem! These three equations aren't enough: Every one of the distances is wrong because the clock in the receiver is not accurate.

We've assumed above that the receiver is synchronized with the satellites, which it would be if it had an atomic clock built into it, but, as noted above, to keep the receiver cheap and compact it has only a low accuracy clock. So, we need to correct the distances by the error in the receiver's clock. If the receiver differed from satellites by Δt then the distance error is $c\Delta t$ where c is the speed of light. Our three equations then become:

$$d_1 + c\Delta t = \sqrt{(x_r - x_1)^2 + (y_r - y_1)^2 + (z_r - z_1)^2}$$
$$d_2 + c\Delta t = \sqrt{(x_r - x_2)^2 + (y_r - y_2)^2 + (z_r - z_2)^2}$$
$$d_3 + c\Delta t = \sqrt{(x_r - x_3)^2 + (y_r - y_3)^2 + (z_r - z_3)^2}$$

We now have three equations and four variables: x_r, y_r and z_r, and now Δt. We need a fourth equation, which we get from a fourth satellite. Just as with the first three satellites, the receiver calculates its position from this fourth satellite:

$$d_4 + c\Delta t = \sqrt{(x_r - x_4)^2 + (y_r - y_4)^2 + (z_r - z_4)^2}$$

We now have four unknowns and four equations. These four equations are highly non-linear and cannot be easily solved in the usual way by substituting. Instead, engineers use a procedure called Newton-Raphson iteration.

To implement this iteration, they expand each equation into an infinitely long polynomial based on the trial set of values for the position. These four linearized equations can be solved simultaneously to determine whether this guess was right; if not, we substitute the values generated and repeat. If the receiver starts with a good guess of where it is—usually its last location—

then this can often converge in one iteration. If you instead used the GPS receiver, turned it off, and then flew half way around the world, it would take a bit of time for the receiver to locate itself.

Primer: Nuclear Structure

IN THE CHAPTER THAT FOLLOWS, we discuss the engineering concepts behind enriching uranium for use as fuel or in a nuclear bomb, but to set up our understanding of this process, we must have an understanding of the structure of atoms in general.

All atoms consist of a nucleus composed of protons and neutrons surrounded by a cloud of negatively charged electrons. The protons are positively charged, while the neutrons have no charge, but they are each 2,000 times the size of an electron. Thus, the mass of an atom is pretty much just the size of its nucleus. Several forms of many atoms occur naturally, and we call these different forms *isotopes*. Isotopes differ from one another in the number of neutrons each isotope contains; for instance, while the most common naturally-occurring form of the element carbon has 6 neutrons, the isotope of carbon that has 6 protons and 8 neutrons in its nucleus (used for carbon dating) is written as ^{14}C, where the superscript lists the number of protons and neutrons together. Uranium, the element of discussion in this chapter, occurs in nature as mostly isotope ^{238}U (~99%), with a small amount of ^{235}U (0.72%).

The nucleus is filled with positively-charged protons, which repel each other. The force that keeps the nucleus from flying apart is known as the strong nuclear force. This is one of four fundamental forces of nature (the other three being the weak nuclear force, electromagnetism, and gravity) and is also the strongest. It is incredibly powerful at short distances, but at distances more than a few angstroms (1 x 10⁻¹⁰ m) long, the force drops to zero. The balance between these two opposing forces—protons repelling each other and the strong nuclear force keeping them together—helps to explain why some elements can undergo fission.

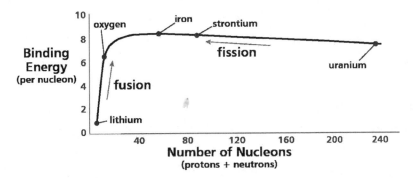

This graph shows the amount of binding energy per nucleon. (A nucleon is either a proton or a neutron.) The x-axis shows the total number of nucleons, which equals the atomic mass. Thus, the elements appear from lightest to heaviest from left to right on the x-axis. The greater the binding energy, the more stable the nucleus. The peak occurs at ~60, which corresponds to iron, the most stable element in the universe. Elements to the left of iron have a tendency to combine (fusion) to become more stable. For example, in the sun, hydrogen, the lightest element, undergoes fusion to become helium, the second lightest element. Elements to the right of iron can become more stable by splitting apart (fission).

Fission

In undergoing fission, a nucleus splits apart, releasing the energy binding it together. Elements with atomic numbers larger than iron are less stable than those with fewer protons. As the atomic number increases, the number of protons being packed together increases, yet geometry indicates that they must become further apart as their number increases. This means that the strong interaction is diminished, yet every pair of protons in the nucleus stills feels an electrical repulsion. Thus, the force that holds the nucleus together weakens, and the force driving it apart increases. This makes atoms of high atomic number susceptible to fission. Nuclei are typically stable, so fission requires adding energy to the nucleus.

Typically, we break up a nucleus by bombarding it with neutrons. That sounds exotic, but it's good enough to think of the bundle of neutrons and protons at the nucleus's core as a rack of billiard balls and the bombarding neutrons as a cue ball. Just as with cue balls, the degree to which the rack breaks up depends on the speed of the cue ball; the analog in nuclear physics is slow and fast neutrons. Bombarding ^{235}U with slow neutrons causes its nucleus to begin to vibrate violently because the kinetic energy of the moving neutron is transferred to the particles in the nucleus. The uranium atom then breaks apart:

$$^{235}U + neutron \rightarrow {^{90}Sr} + fission\ fragments$$
$$+\ 2\ or\ 3\ neutrons + energy$$

The fragments can be many things, but typically ^{235}U fissions into Strontium-90 (^{90}Sr). Not all nuclei will easily undergo fission. Most elements are stable so that when bombarded with neutrons they simply absorb the neutron and decay later, or they require very high energy (fast) neutrons. To extend the analogy with billiard balls a bit, imagine the rack of balls being "sticky" so they require a faster (higher kinetic energy) cue ball to break them apart. To extract nuclear energy in a practical way requires elements that are reasonably stable so they can be stored without appreciable decay but are capable of undergoing fission with neutrons of all energies. Only four elements meet these criteria: ^{235}U, ^{233}U and the plutonium isotopes ^{239}Pu and ^{240}Pu.

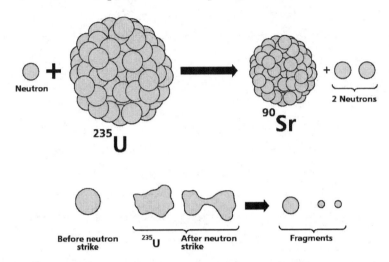

You can visualize fission by picturing the uranium atom as a spherical liquid drop before the neutron hits; after impact, it vibrates, becoming elliptical and stretching until it looks like a Q-tip. Eventually the thin section breaks and the nucleus splits—fissions—into smaller parts, with splinters flying off. For ^{235}U those splinters are typically two or three neutrons.

Nuclear Explosions & Chain Reactions

To make a nuclear reaction that generates a great quantity of energy requires a chain reaction that releases energy as it proceeds. For the chaotic explosion of a bomb the reaction releases more and more energy; where as in a nuclear power plant the reaction releases enough energy to just sustain the reaction. Here ^{235}U has the two essential properties to sustain a chain reaction. First, it can undergo fission from slow neutrons and, second, that fission generates more than one neutron. To see why both of these are needed, let's look at trying to generate a chain reaction in natural uranium.

Natural uranium is made up of mostly ^{238}U with a tiny bit of ^{235}U. Fast neutrons will cause both ^{238}U and ^{235}U to fission. The neutrons created when either of these isotopes fission, however, will be slow neutrons. Slow neutrons will cause ^{235}U to fission, but not ^{238}U. Instead, a ^{238}U atom absorbs slow neutrons and then decays later without releasing any neutrons that would continue the chain reaction. Effectively the ^{238}U quenches the chain reaction. The case is even worse if we start with slow neutrons: Only 1 in 140 ^{235}U atoms will fission, and ^{238}U never will. To make ^{235}U undergo a chain reaction, we need to remove some of the ^{238}U. This process of creating uranium with a ^{235}U concentration greater than found in nature is called enriching. Typically when natural uranium is enriched to 3 or 4% ^{235}U, it can fuel nuclear reactors; enriching to 90% powers a nuclear bomb.

Uranium
[yoo-rayn-i-uhm] | Symbol: U
Atomic number: 92 | Atomic weight: 238.0289 amu

Named for the planet Uranus, this element's defining characteristic lies in its nearly unique ability to start a nuclear chain reaction that releases tremendous amounts of energy. One kilogram of uranium has the energy of over 1,000 tons of TNT. *Splitting the uranium atom offers the promise of bountiful energy via nuclear power, but also the ability to destroy the world with atomic bombs. Uranium, a silvery, ductile metal, powered the first nuclear bomb. On August 6, 1945, at 8:16* AM, *the United States dropped a bomb codenamed Little Boy on Hiroshima, Japan. Uranium brought massive changes to our society, allowing it to compete with silicon to define the twentieth century. Future historians will call the century the dawn of electronics, or the opening of a century of death, depending on what plays out in the twenty-first century. Oddly, for an element with such life-changing promise or peril, it isn't rare; it's more common in the earth's crust than tin. What prevents nuclear weapons from proliferating around the globe is the difficulty with purifying uranium so it can power a bomb.*

The Hardest Step in Making a Nuclear Bomb

I DON'T BELIEVE A WORD of the whole thing," said Werner Heisenberg, on hearing of the nuclear bomb dropped on Hiroshima. Allied forces had captured Heisenberg, winner of the 1932 Nobel Prize in Physics for his uncertainty principle, as part of an operation begun on February 24, 1945. The Allies detained Heisenberg and nine other top German nuclear scientists, including fellow Nobel Laureates Otto Hahn and Max von Laue. They kept them secreted away at a large, isolated house fifteen miles from Cambridge, where the British secret service clandestinely recorded their conversations. Shortly before dinner on August 6, 1945, Otto Hahn told the others of a BBC announcement that an atomic bomb had been dropped on Japan. Immediately Hahn added, "They can have only done that if they have uranium isotope separation." Heisenberg, doubtful of the report, chimed in that the United States' Manhattan Project "must have spent the whole of their £500,000,000 in separating isotopes." Hahn felt separating isotopes was still twenty years away; another scientist added that he didn't think the bomb had anything to do with uranium. The scientists continued discussing the difficulty and outlined methods of separating isotopes. It's telling that of all the steps to make a nuclear bomb, these experts focused on one aspect: creating the proper type of uranium for a bomb.

Operation Alsos: How We Know What Heisenberg Said

An Allied raiding party snuck through a gap in Nazi Germany's crumbling front. As part of Operation Alsos—a largely British-United States effort to find out just how close the Nazis were to a nuclear bomb—the soldiers were to seize any German nuclear resources. The team pushed through the Eastern edge of the Black Forest to find their target: the Kaiser Wilhelm Instiut fur Physics, the site of most of Germany's nuclear research. They captured Nobel Laureates Otto Hahn and Max von Laue and learned that Werner Heisenberg, also a Nobel Prize winner, had escaped two weeks before. Under interrogation, the scientists revealed that Germany's most secret nuclear information was sealed in a metal drum sunk in a cesspool behind the house of physicist Carl von Weizsacker—a man famed in the early 1930s for explaining the nuclear process inside stars. With this information, the allied forces soon captured the remaining chief scientists, including Heisenberg. The Allies faced a problem: If they formally arrested and charged these scientists with war crimes, they would call attention to the two US nuclear bombs being readied for testing in America. One general suggested that the simplest solution would be to shoot all the scientists. Unpalatable and barbaric to the Allied Command, they eventually decided to keep them detained incommunicado at an isolated house fifteen miles from Cambridge. Owned by the British secret service, the house, called Farm Hill, had microphones installed, a standard practice with senior prisoners of war, although the detained scientists themselves were unaware of the microphones. One scientist asked Heisenberg, "I wonder whether there are microphones installed here?" Heisenberg replied, "Microphones installed? (Laughter) Oh, no, they're not as cute as all that. I don't think they know the real Gestapo methods; they're a bit old-fashioned in that respect."

Eight Amazing Engineering Stories

Making "highly enriched uranium, the essential ingredient of a nuclear bomb," a high-ranking US official testified to Congress, "is the hardest step in making the weapon." A bomb maker or power plant operator needs a rare variant of the element uranium, the isotope ^{235}U, that easily releases its nuclear binding energy. While the principles behind a nuclear bomb can be understood by anyone with basic scientific knowledge, the engineering to enrich uranium is tremendously difficult since nature doesn't make it easy. The key engineering problem lies in separating two nearly identical uranium isotopes.

To extract nuclear energy in a practical way requires elements that are reasonably stable so they can be stored without appreciably decaying, but that are capable of undergoing fission with neutrons of all energies. Only four elements meet these criteria: ^{235}U, ^{233}U and two plutonium isotopes ^{239}Pu and ^{241}Pu. Of these, ^{233}U is even rarer than ^{235}U, and plutonium is made in a nuclear reactor from ^{238}U, so in some sense is harder to produce than ^{235}U. Pure ^{235}U can easily sustain a chain reaction that releases tremendous energy. Each time a ^{235}U atom undergoes fission, it releases a tiny amount of energy, roughly three million times the energy released when a single octane molecule in gasoline combusts. The key, of course, is that this happens very many times and, because of the chain reaction, it occurs at nearly the same time. This chain reaction cannot happen in naturally occurring uranium, which consists mostly of ^{238}U; only 1 atom in 140 is ^{235}U. The ^{238}U will quench completely the chain reaction. So, to get such destructive power, a bomb maker needs to remove the ^{238}U in natural uranium so it becomes enriched in ^{235}U. For uranium to be used as fuel for nuclear power, it must be enriched to 4% ^{235}U. However, enriching natural uranium to 90% ^{235}U provides enough of the fissile ^{235}U to power a nuclear bomb. Luckily this is a difficult engineering problem to solve.

> **The Power of the First Nuclear Bomb**
>
> The first nuclear bomb, which destroyed Hiroshima, contained about 60 kilograms (about 132 pounds) of uranium ^{235}U, of which only 600 g (1.33 pounds)—or about 100 sextillion ^{235}U atoms—underwent fission. This was enough, though, to create an explosion equal to 13 to 18 kilotons of TNT.

Why It's Hard to Enrich Uranium

When we separate two items, we make use of their differences. For example, to sort silverware we note the different shapes of spoons, forks, and knives. Industrially, the same principle applies: To separate iron from plastic in a recycling center, we use a magnet; copper is leeched from its ore by sulfuric acid; and to desalinate ocean water, engineers boil the water, producing steam consisting of pure water. However, the two uranium isotopes ^{235}U and ^{238}U have nearly identical physical, magnetic, and chemical properties. This means that no magnets will tug on one more than the other, no solvent will wash away only one isotope, and neither will boil before the other. So, to separate them, engineers exploit the one slight difference between them: ^{235}U weighs slightly less than ^{238}U. That tiny weight difference means the two isotopes will move at slightly different speeds when exposed to a force. With less than a two percent difference, it's just enough to make separation possible, but not easy. Consider this: If these two uranium isotopes were to race the 2,500 miles from New

York to Los Angeles, the lighter isotope would arrive 25 feet ahead of the heavier one.

How Uranium Was Enriched for the First Atomic Bomb

To enrich uranium for the first atomic bomb, engineers built immense gaseous diffusion plants that capitalized on the fact that ^{238}U would move slower. In essence, these plants allow a gas of uranium to flow through miles of piping in a kind of race, where at the end the lighter ^{235}U wins out. Such plants are immense: For the Manhattan Project, which built the first atomic bomb, the building housing the diffusion plant covered over 40 acres and was a half-mile long in the shape of a "u," containing a twisted maze of 100 miles of piping. To start this "race," engineers would combine uranium with fluorine to create uranium hexafluoride—$^{235}UF_6$ and $^{238}UF_6$. Although solid at room temperature, UF_6 turns into a gas, commonly called "hex," at 56.5 °C. Next, they sent this gas through a tube encased in a chamber.

Lower pressure in the chamber compared to the tube results in more of the $^{235}UF_6$ than $^{238}UF_6$ passing through perforations in the tube's wall. The chamber traps any gas that escapes the tube. As you might guess, the amount of separation, or enrichment, is slight: An entering stream of hex containing 3% ^{235}U would exit only a tiny bit richer in ^{235}U—perhaps ~3.013%. So, to increase the separation, this slightly enriched stream is passed through another tube and chamber, called a stage, which enriches it a bit more, and then through another stage and another. To enrich a stream of 3% ^{235}U to 90% takes nearly 4,000 of these stages.

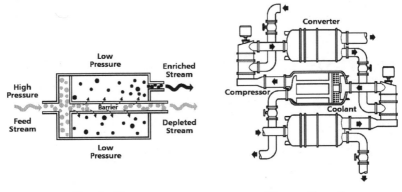

Single Stage   **Stages Together**

In the single stage shown on the left, *a pressure difference forces hex gas containing both ^{235}U and ^{238}U isotopes to flow through a porous pipe. The light ^{235}U travels faster and further and thus is more likely to pass through the barrier. The isotope separation in a single stage is very small. To achieve great separation, many such stages are linked together. The drawing on the right shows three linked stages. Note that the enriched stream passes into the next stage. Thus this stream becomes more and more enriched as it passes through the stages.*

Gaseous Diffusion

A gaseous diffusion plant, like the one used in the Manhattan Project, works by molecular effusion, meaning that one of the two isotopes to be separated is more likely to pass through the perforations in the tube. This tube is, more precisely, called a barrier, but these perforations are not microscopic. They are sized based on results from the kinetic theory of gases that show molecules travel at different speeds because of differences in their masses. Specifically,

$$v_{rms} = \sqrt{\frac{3kT}{m}}$$

where v_{rms} is the root mean square velocity (a type of average velocity), k is Boltzmann's constant, T is the absolute temperature, and m is the mass of the particle.

Because the velocity is inversely proportional to the square root of the mass, a lighter particle will travel faster than a heavier one. This means that in a gas mixture of two particles of different weights, the lighter one will have a higher probability of hitting a hole in the barrier. It has a higher probability because it travels further on average than does a heavier molecule; this typical distance traveled, called the mean free path, sets the size of the holes in the barrier.

At atmospheric pressure, the mean free path of a molecule is about one ten-thousandth of a millimeter or one-tenth of a micron. This is just another way of representing the distance that, on average, a molecule travels before colliding with another molecule that changes its direction or energy. Thus, to ensure the necessary separation for an atomic bomb, the diameter of the billions of holes in the barrier must be less than one-tenth the mean free path, otherwise both effusing isotopes would pass through the barrier. The exact composition and preparation of the barriers used today in separating uranium are secret, but for earlier barriers, engineers rolled Teflon powder onto nickel gauze to create pores of 0.015 microns.

Modern Method: Centrifuge

Gaseous diffusion plants are very large and thus expensive to build and operate. Part of this cost is that they take great amounts of energy to run. For example, all of the compressors and heaters generating pressure and heating the hex (the uranium fluoride gas) throughout the miles of tubing require vast amounts of energy to operate. Early plants used so much energy that many people wondered whether it would take more energy to enrich the uranium than the energy produced by that uranium in a nuclear reactor! At roughly the same time as the first gaseous diffusion plants were developed, a few scientists and engineers realized that they could exploit the small mass difference using a centrifuge. A centrifuge spinning at thousands of revolutions per second will force more of the heavier ^{238}UF$_6$ to its outer wall than ^{235}UF$_6$. Typically, a gas centrifuge used to separate uranium is called a rotor, which is a long thin tube.

Hex gas, which is a mixture of the two isotopes, flows in at the top of the tube and travels down to the center, where it's released into the rotor. As the rotor spins around its long axis, more of the heavier ^{238}U is thrown out to the edge than the lighter ^{235}U, creating two layers of gases. The first layer is pressed up against the outside of the rotor and contains a higher percentage of the heavier ^{238}U, while the layer closer to the center contains a bit more ^{235}U than in the gas near the wall. However, the centrifuge compresses nearly all the gas in the rotor to the wall, meaning that most of the distance from the center to the outer edge is empty. This in turn creates a higher pressure at the outer wall than near the interior, since there are more gas particles being pushed to that outside wall. The faster the rotor spins, the greater the separation. However, the separation is always tiny: For a single rotor, both layers near the wall contain a lot of ^{238}U. To increase the separation, engineers use one more trick.

By heating the bottom of the rotor, engineers induce a flow of gas. The heating expands the gas slightly, lowering its density. The lower density gas moves toward the center of the rotor where the pressure is lower. So, the gas at the outer wall flows down as the gas nearer to the center flows up. This counter-current flow increases the amount of separation. As the gas travels up the rotor, its rotation depletes more and more of the heavier ^{238}U. At the top, engineers remove the slightly enriched stream, and from

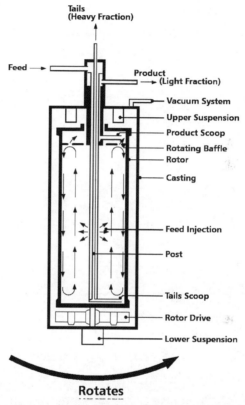

In a typical gas centrifuge, hex gas enters at the top and flows through a tube on the centrifuge's vertical axis. The feed enters halfway down the tube. As the centrifuge spins, more of the heavier isotope—^{238}U—is thrown toward the outer wall than the lighter isotope—^{235}U. A small scoop at the bottom heats the gas by friction—the gas moves extremely fast over the scoop. This temperature difference between top and bottom creates a flow and increases the amount of separation. The "heavy" fraction—the stream richest in ^{238}U—is drawn by a tube from the bottom of the centrifuge; the "light" fraction—the stream richest in ^{235}U—is drawn from the top.

the bottom withdraw the slightly depleted stream. This type of flow increases the separation by a factor of two over just spinning the rotor. For a typical single centrifuge with natural uranium entering at 0.71% ^{235}U, the exiting stream will have 0.79% ^{235}U. Still tiny, but greater than for gas diffusion. Yet, just as with the gas diffusion plants, we need to use stage after stage to fully enrich the uranium. A typical gas centrifuge plant might have about 60,000 centrifuges to enrich natural uranium to 3% ^{235}U. Such a plant uses 1/25 the energy of a gaseous diffusion plant.

This all sounds simple enough, so why isn't the world filled with a nuclear weapon in every home that desires one? It's simple: The engineering of a gas centrifuge plant is very difficult.

Why It's Hard to Make a Rotor

No engineer can make a perfectly balanced rotor. By that, we mean that its geometric center, the long axis of a cylinder, coincides perfectly with its center of mass. Think of a rigid cylinder with a large lump attached to its outer wall: When it spins, it will wobble as the asymmetrically placed weight tugs on the wall. Engineers make the rigid rotors of a gas centrifuge with great precision, but there are small variations in the wall thickness that make it an imperfect cylinder. These variations function just like the large mass attached to the outside of the cylinder. While they seem slight, the speed of rotation makes great demands on the precision needed. The force from the imperfections in the cylinder's wall—the extra mass here and there—increases with the square of the rotation speed. For example, a rotor spinning at 1,000 rev/sec with a wall that is 25 microns (1×10^{-3} inches) thicker in spots will tug on the wall with a force equal to ten times the rotor's own weight! This puts an upper limit on the rotational speed of the rotor: A 5-inch diameter aluminum rotor will burst its walls with a little over 1,000 revolutions per second. So, the fundamental problem or trade-off for any engineer is designing a rotor that will spin as fast as possible to achieve maximum separation, and thus reduce the number of stages, yet will not spin so fast that the rotor falls apart. The maximum speed puts a theoretical cap on the separation, but problems occur long before this point.

As the rotor becomes longer and narrower, it will bend at specific rotation speeds. Called "flexural modes," they happen in a particular order and at a rotational speed that depends on the dimensions of the rotor.

A 5-inch diameter rotor that is about 9 inches long will bend into an arc shape at about 180 revolutions per second, then into an "s" at 500 revolutions/second, and a "w" at 1,000 revolutions/second. Operating the rotor at any of these critical speeds where flexure occurs will cause the rotor to blow itself apart. Therein lies an engineering dilemma: If you operate the rotor below the first critical speed, it will spin too slowly to create effective separation, but spinning faster for a greater separation could destroy the rotor.

The first gas centrifuges resolved this dilemma by accelerating the rotor rapidly up to operating speed, passing through the dangerous speeds for only a fraction of a second. (A rotor typically operates just below one of the higher critical speeds.)

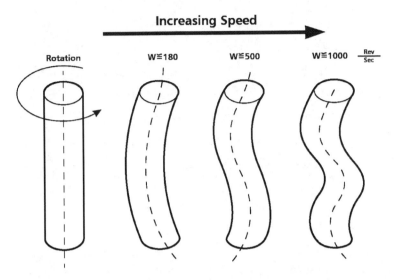

As the rotational speed of a rotor increases, it begins to flex. Called "flexural critical speeds," they can be alleviated by making a short tube, but a short tube creates a smaller degree of separation.

Although simple, this method uses a great deal of energy. Today's engineers use a more energy-efficient approach: Modern centrifuges use a clever design to reduce the damage done by the critical frequencies.

It would seem that the best way to keep a rotor from vibrating would be to use extremely tight bearings on each end. Yet the opposite works better: Engineers affix the rotor to the bearing using flexible shafts that allow the rotor to find its own center of rotation, its natural center of gravity. This isn't the center of the rotor's cylinder, and so the rotor rotates a bit askew. Such a rotor spins with an almost uncanny silence and with no vibration. While flexible bearings help to ameliorate the effects of imperfections in the rotor's wall thickness, they cause problems at the critical flexural speeds because the rotor can now shake even more violently. To compensate for this increase in deformation, engineers attach springs to the bearings, weighted so that when

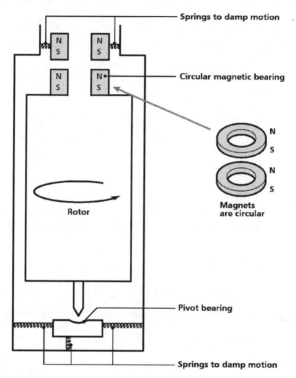

A pair of magnets restricts the motion of the top end of the cylinder so that it doesn't topple. The bottom magnet can spin, kept in place by the top magnet. Unlike most bearings, this magnet has low stiffness: The rotor can shift off axis a bit and still be held in place by the magnetic force. This works like a flexible shaft that lets the rotor find its own center of rotation. In addition to their flexibility, magnetic bearings have low friction, so they don't dissipate much energy.

the critical speeds arrive, they absorb or dampen the vibrations. Modern gas centrifuges don't use flexible physical shafts, but instead an ingenious magnetic bearing.

A typical gas centrifuge sits on a metal needle in a sapphire cup, called a pivot bearing, which allows the cylinder to spin like a top. As described in the drawing, a set of magnets allow the bearing to spin freely. These bearings have a low stiffness and so allow the rotor to shift off axis a bit and still be held in place. In addition to their flexibility, magnetic bearings have low friction, so they don't dissipate much energy.

However, magnets' big disadvantage is that the rotor is more delicate and susceptible to disturbances. Of course, this difficulty is where we began: The engineering and operation of a gas centrifuge isn't easy! This makes building a uranium separation plant difficult and allows tracking of the countries clandestinely installing these plants. Making gas centrifuges to enrich uranium is so specialized that to track the spread of nuclear weapons and to prevent their proliferation, the International Atomic Energy Agency, an independent agency that reports to the United Nations General Assembly, tracks specific materials that could be components of an enrichment plant. On their "track" list are thin-walled tubing of high strength, powerful ring-shaped magnets, and high-speed multiphase motors. All of these are components necessary, of course, to create a hollow rotor spinning at high speeds.

The Future

As with any technological problem, engineers focus strongly on its Achilles Heel, its hardest, most expensive step, in order to improve the technology. For producing nuclear energy or a nuclear weapon, as we've just seen, it's the enriching of the uranium so that it will easily undergo a chain reaction and release energy. The two methods currently used, gaseous diffusion and centrifuges, have their roots in the early 1940s. It's no surprise, then, that since that time, engineers have developed ways to solve the perplexing problem of uranium enrichment.

Atomic vapor laser isotope separation (AVLIS) exploits the tiny difference in the frequency (or color) of light absorbed by ^{235}U and ^{238}U. Although these frequencies differ by only one part in a million, the lasers used in AVLIS can be tuned so that only ^{235}U atoms absorb the laser light. When it absorbs the light, ^{235}U will eject an electron and become a positively charged ion. Engineers can then use an electrostatic field to separate the two isotopes: Charged ^{235}U atoms go one way, and neutral ^{238}U atoms the other.

Another process, molecular laser isotope separation (MLIS) uses two basic steps. First, engineers irradiate UF_6 with 16 mm wavelength light from an infrared laser, which only excites the $^{235}UF_6$. Second, photons from a second laser (infrared or ultraviolet) preferentially dissociate the excited $^{235}UF_6$ to form $^{235}UF_5$ and free fluorine atoms. The $^{235}UF_5$ precipitates from the gas as a powder that can be filtered easily from the gas stream.

No doubt we should be happy that an engineering barrier exists between the world and terrorists, but we should never forget that all technological solutions for quelling terrorism are at best a stopgap measure. We all want some quick technological fix to the problems of terrorism: an ultra-sensitive metal detector that will unmask any weapon or a device to detect instantly the production of enriched uranium. Yet, the only lasting solution will be a human one: an alert person at an airport, trained law enforcement personnel, or agreement and understanding among the peoples of the world.

Lead
[led] | Symbol: Pb
Atomic number: 82 | Atomic weight: 207.19 amu

A freshly-cut piece of lead sparkles with a blue-white sheen, but within a few minutes it oxidizes to the gray color you are probably familiar with. Perhaps no other element's properties are as recognizable as lead. Who cannot picture, for example, its great density—think for a moment of the astounding heaviness of a car battery. Although somewhat scarce, it can be found pure in nature, and so became one of the first metals used and named by humans. It is the only element mentioned by name in the Bible; Exodus describes the Red Sea engulfing Pharaoh's chariots as "the sea covered them: they sank as lead in the mighty waters." Its symbol, Pb, comes from the Latin plumbum, *meaning lead; it forms the root of our modern day* plumber *because the Romans used lead in the pipes that supplied their cities with water. Just as with the Romans, lead still forms an everyday and essential part of our world: Many cars, trucks, and planes start only because of the irreplaceable lead-acid battery.*

The Lead-Acid Battery: A Nineteenth Century Invention for the Twenty-First Century

BATTERIES LIE AT THE CENTER of our technological world. Not only do they power cell phones, flash lights, and laptops, but more importantly they make our world move: The single largest use of batteries is for starting the engines of cars, trucks, and some jets. And batteries will only become more important as we continue to move from fossil fuels to alternative sources like wind and solar that require batteries to store their energy.

Yet innovations in batteries are infrequent and hard won. Over the last twenty years, the amount of energy stored in a typical laptop battery has increased by less than 10% per year, while in the same period the number of transistors on a microprocessor changed from one million to over one billion.

The commonplace lead-acid battery in a car highlights the incremental advances made in batteries. Introduced in 1860, the lead-acid battery is still used to start our cars. While this seems like a victory for nineteenth century technology, in reality it's a loss for our era. There is nothing car makers and jet manufacturers would like more than to replace the heavy, chemical-filled, lead-acid battery with something lighter in weight and less toxic. The fact that we currently still use the clunky lead-acid battery reveals a great deal about battery design.

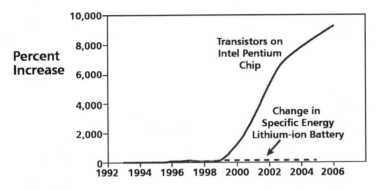

Since 1993, the number of transistors on an Intel-brand computer CPU has increased from 3,100,000 in 1993 on their first Pentium chip to 291,000,000 on their 2006 Core 2 Duo. In contrast, the energy density of a lithium-ion battery has increased over the same time period from 89 to a little over 200 watt hours (Wh)/kg.

Are Electric Cars Really Better for the Environment?

A focal point in the discussion of climate change is the emission of greenhouse gases, especially the contribution of carbon dioxide to the warming of the atmosphere. A popular "green" choice to alleviate some of the environmental impacts of fossil fuel-powered cars is the use of electric or hybrid electric cars. However, the case for such cars isn't as clear as you would expect. The proper way to assess the impact of an object on the environment is life-cycle analysis; that is, looking at its environmental impact at every stage of its life, from birth to death.

Several researchers have compared the impact of traditional gasoline-fueled vehicles, electric cars, and hybrids like the Toyota Prius over their lifetimes. These studies show that while electric cars produce far fewer emissions throughout their performing life, they take an incredible amount of energy to produce. *Wired Magazine* reported that the Toyota Prius consumes over 1,000 gallons of gas before it even drives its first mile! The nickel batteries that power the car require lots of fossil fuel energy to come to life. The nickel is mined in Canada, travels across the world to Europe for processing, and then on to China for further processing. From there, it is transported to Japan where the cars are manufactured, and then finally to the United States to be sold. These trips, as well as the processing taking place at each stop, release emissions that begin to rack up before the car even hits the dealership. Factor into this equation the small operating life of the battery (typically no more than 100,000 miles), and the electric car doesn't look as green as it once did. Ultimately, it's tough to say whether or not the electric or hybrid car outperforms a new gas car, but it is likely that a fuel-efficient used car is better right now than either option because the first owner has already paid off the carbon debt of the car's manufacture. And with dozens of cars from the 1990's that top out at over 30 miles per gallon, you have a decent selection to choose from. The overall lesson here? Like many of today's problems, there is no perfect solution available yet.

Basis of All Batteries: Transfer of Electrons

The basis of all batteries lies in the transfer of electrons between two chemical species to create electricity, specifically, to produce a current by the flow of electrons. The role of the engineer in designing the battery is to create a way to extract useful energy from that electron transfer. The lead-acid battery stores and releases energy by the exchange of an electron between pure lead and lead oxide in sulfuric acid. The lead gives up an electron, which the lead oxide accepts. These reactions will occur just by mixing pellets of pure lead (Pb) and pure lead oxide

(PbO₂) in a sulfuric acid-water solution. This exchange turns both into solid lead sulfate, which precipitates as follows:

$$Pb_{(s)} + 2H_2SO_4 \text{ (in solution)} \rightarrow PbSO_{4(s)} + 2H^+ \text{(in solution)} + 2e^-$$

$$PbO_{2(s)} + 2H^+ \text{(in solution)} + 2H_2SO_4 \text{ (in solution)} + 2e^- \rightarrow$$

$$PbSO_{4\;(s)} + 2H_2O$$

But these reactions will yield no useful current because the electron transfers happen all over the beaker, so the net current is zero (recall that current, here, is the flow of electrons). But if we set things up in the right way, we can make a battery; that is, we can get that electron to flow through a wire and thus generate electricity.

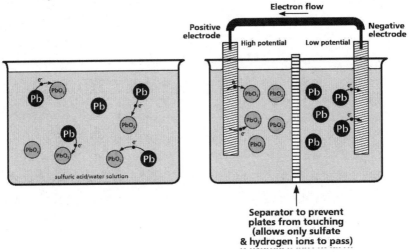

The electrochemical reactions on the left *generate no useful electricity from the exchange of an electron between lead and lead oxide. The net current (flow of electrons) is zero. To make a battery, an engineer needs to arrange things so the electrons flow through a wire. As shown on the* right, *the basic components of a battery are: Positive and negative electrodes linked by a wire outside of the solution, an electrolyte solution that can transport charge between the electrodes using ions, and a separator that allows the ions to pass but keeps the electrodes separated.*

Elements of a Battery

To get useful work from the conversion of chemical energy into electrical energy, we need to arrange the constituents above in a special way. Specifically, we shape the lead and lead oxide into separate solid electrodes and keep them separated. If we place these two electrodes into a water/sulfuric acid solution and connect them with a wire, we find that a current will flow along the wire. The ions H^+ and SO_4^{2-} carry a charge between the two electrodes, while the electron that needs to be exchanged between the lead and lead sulfate electrodes is carried through the wire. This is all fine in principle, but it doesn't make a practical battery that can be sold.

> **Electrolytes**
>
> We call the sulfuric acid/water solution the electrolyte. An electrolyte is any substance that has free ions that make the substance electrically conductive. Typically, this is an ionic solution, but electrolytes can be made from molten salts and even solids. For example, negatively-charged oxygen atoms in solid yttria-stabilized zirconia, a zirconium-oxide based ceramic, can act as charge carriers.

Engineering a Useful Battery

If we just put two electrodes in a solution, the current generated will be very, very low. To make a useful battery, we need to produce a substantial current. For example, a car battery delivers a short burst of 400 to 600 amps, while our previous set-up can only generate a few milliamps—one hundred thousandth of the current needed. To create the current needed to start a car, we need to have electrodes with large surface areas. Large, flat electrodes allow many electrons to transfer at once, helping create a significant current. To make a useful battery, not only do the electrodes need high surface areas, but they must also be packed into the battery as tightly as possible to increase the energy density. Herein lies the difficulty in making a useful battery. Although we want to pack many large and flat electrodes in a battery, if the electrodes inside a battery touch, the produced current will diminish. To prevent this from happening in a commercial battery, we add a material that separates the electrodes from each other and acts as an insulator but still allows ions to pass through.

A typical car battery is composed of six individual cells in series. Each of the cells generates around 2 volts; thus six of them create the typical 12-volt battery used in cars and motorcycles. Within a cell are thin, flat lead and lead oxide plates. Because lead is soft and malleable, or "not mechanically stable" in engineering terms, engineers typically combine it with antimony or calcium. Often, there will be three lead electrode plates alternating with three lead oxide plates, which are kept apart by a separator. The micro-porous separators, made of paper-like materials in cheaper batteries or more sophisticated glass mat in more expensive units, keep the two types of electrodes from touching (and thus shorting out), yet allow sulfate and hydrogen ions to move between the electrode plates.

Eight Amazing Engineering Stories

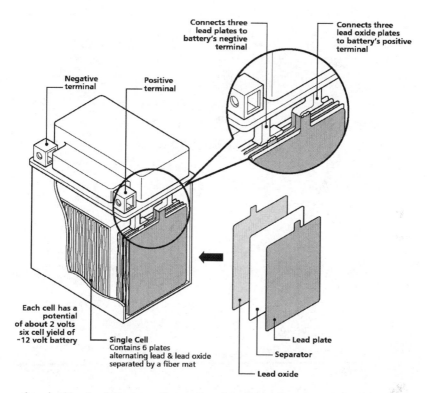

A typical lead-acid battery: It contains six cells, each composed of alternating lead and lead oxide plates separated by materials that keep the plates from touching, but let the ions in the sulfuric acid/water solution pass.

When a battery discharges, it releases electrons at the negative electrode, which then flow through the load to the positive terminal where they enter into a reaction with the lead oxide electrode. As this reaction proceeds, the sulfate ions in the electrolyte deposit onto each of the two electrodes. This lead sulfate, which is an electrical insulator, blankets the electrodes, leaving less and less active area for the reactions to take place. As the battery becomes fully discharged, the cell voltage drops sharply and its internal resistance rises abruptly because of the quantity of lead sulfate on the electrodes.

> ### The First Battery?
>
> In 1936, Wilhelm König, an Austrian painter, was studying the artifacts stored at the Baghdad Museum when he came across an object that intrigued him: a simple ovoid ceramic jar about 14 cm tall and 8 cm in diameter. It had been uncovered by the Iraq Antiquities Department while excavating an ancient Persian settlement and was dated to between the first century BC and the first century AD. Inside the jar, König noticed a coiled copper sheet affixed to the walls by asphalt. The tube was closed at the bottom and covered with a thin layer of asphalt. Inside the hollow tube hung an iron rod, suspended by an asphalt plug at the top; it was concentric with the copper cylinder's axis so that nowhere did it touch the copper. The asphalt seal suggested to him that this arrangement could hold liquid. The presence of dissimilar metals in an acid solution—perhaps freshly squeezed grapefruit juice—generates a potential difference. To König, the contraption resembled a wet-cell—a galvanic cell with a liquid electrolyte. He suggested it was used for electroplating. While it would seem impossible to have been built based on some ancient theory of electrochemistry, the effects of two dissimilar metals could have been observed. Use a bronze spoon in an iron bowl containing vinegar, and you will feel a tingling in your hand. No one knows exactly what this device is, or what it was used for, but it offers the intriguing possibility of an early battery.

Why Discharged Car Batteries Can Freeze

The number of free sulfate ions in the acid/water solution provides an excellent measure of a battery's charge. As the sulfate ions "plate out" on to the electrode, the density of the electrolyte drops. This makes the battery vulnerable to freezing when discharged, since the dissolved sulfate ions lower the freezing point of the electrolyte. For instance, while a completely charged battery freezes at -71 °F, a fully discharged one freezes at 17 °F. In an actual lead-acid battery, all of the sulfate ions are never completely plated out of solution before the battery fails to deliver current. At full charge, there would be about 30% sulfuric acid by weight, and about 10% when completely discharged.

When the battery is recharged, the opposite reactions occur. As electricity is supplied to the battery, the battery's voltage rises and the internal resistance drops. These changes occur because the supplied current drives a reverse reaction that removes lead sulfate from the plates, causing it to return to the liquid electrolyte. Understanding the movement of the sulfate ions from liquid to solid and back again allows us to understand why batteries eventually wear out. During recharge, not all of the lead sulfate on the electrodes returns to solution. Over many charge-discharge cycles, this accumulation becomes great enough that the resistance of the battery rises and it can no longer be recharged. The amount of lead sulfate permanently bound to the electrodes increases the longer it's allowed to stay on the electrodes, so it pays to keep a battery as fully charged as possible. In the large current surge that starts a car, about 20% of the battery's charge is used up. After starting, the car's alternator immediately starts to recharge the battery to 100% as the engine runs. This is why your battery dies if your alternator fails: Every start uses up some charge without replacing it.

> **Anodes and Cathodes**
>
> Electrochemical engineers have moved away from designating electrodes as anodes or cathodes and instead label the battery's terminals as negative or positive. They do this because this designation never changes during charging or discharging. For a lead-acid battery, the lead electrode is always the negative electrode and lead oxide the positive one, while the electrode that acts as the anode during discharging becomes a cathode during charging, and vice versa.

Why the Perfect Battery Doesn't Exist

There's an old engineering joke that we like: "If you want something good, fast, and cheap, you can only choose two." This applies, in a way, to batteries. Every battery is a trade-off, by which we mean that there are multiple desirable characteristics that we want, but we simply cannot get them all in a single battery, so we find a balance among them. This means that we choose and design a particular battery based on its eventual use. For example, let's contrast batteries used to start a car and those used to power a car and see how differently they are designed. To understand this, it's essential to know the difference between power and energy.

Energy is the capacity to do work. For example, lift a weight in the air, and it has a potential energy that could be converted to work, e.g., by dropping it on something. Energy density is only the amount of energy per unit weight, while power is the rate at which energy can be expended or work done. Just because an object has high energy density doesn't mean it can generate high power. Imagine again that weight on a pulley; let go of the rope, and the weight will drop to the ground quickly. But what if the pulley were rusty so that the pulley turned very slowly? The weight would drop slowly and do much less damage to anything under it. The rusty pulley-weight system has much less power than a well-lubricated pulley-weight system because the rusty pulley-weight cannot expend its stored energy as quickly. A battery dissipates its energy as current, where current is a measure of energy per unit time. So a high current yields high power, while a lower current yields lower power. In a battery, the "rate" of energy flow (the current) depends on the resistance. High resistance gives a small current flow, and low resistance allows for a high flow of current and thus a high-powered battery. In this way, energy density and power density are separate things; a high-energy battery can yield a long-lasting current, and a high-powered battery can yield a very high current for, depending on its size, a shorter amount of time.

The automotive lead-acid battery is optimized for starting, lighting, and ignition (SLI). These applications consume a large amount of energy in short, powerful bursts of current. When starting a car, for example, the current jolt from the battery uses up about 20% of the stored energy before it is recharged. A battery used for a car is designed to never be discharged below about 80%. In fact, the battery will fail after only a few complete charge-discharge cycles. Such a battery would be a poor choice for anything but lighting or ignition.

However, powering a car, rather than just starting it, requires a battery than can be completely discharged and then recharged multiple times. The same goes for a solar energy system: The energy gathered from the sun by the solar panels must be stored in batteries, which a house's electrical system draws on until they are nearly drained. The difference in how you build these two types of batteries highlights the trade-offs in any battery design.

Shallow Charge vs. Deep Discharge

To create an SLI lead-acid battery, also called a shallow-charge battery, requires electrodes that are very thin so that the resistance is low. Low resistance gives us a high-powered battery, since the power of a battery is given by the equation: Power = V^2/R. Here we can see that the power of a battery is inversely proportional to its resistance. Clearly, as resistance drops to zero, the power rises toward infinity. The low-resistance of an SLI or automotive lead-acid battery lies partly in the thin electrodes that decrease the resistance, but this design causes great problems if the battery is fully discharged. When a lead acid battery discharges, it creates insulating lead sulfate, which fills the gaps between the closely-spaced electrodes. At 80% of full charge, there is still enough uncovered electrode area that, when the battery is attached to a power source, current can flow and the lead sulfate can be dissolved back into the liquid electrolyte, thus recharging the battery. However, when the battery is completely discharged, the spaces between the electrodes are completely packed with insulating lead sulfate so that no current can flow and the reaction doesn't reverse. This low-resistance model works great for a device like a car, which is being constantly charged and discharged, but creating a battery to power an electric car or to store solar-generated energy requires a battery that can be fully discharged.

Eight Amazing Engineering Stories

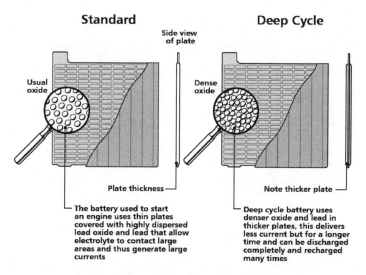

On the left *is a shallow discharge battery, like the one used to start a car. To create large currents, it has very thin plates with highly dispersed lead and lead oxide. In contrast, a deep cycle battery, as shown on the* right, *has thicker plates and more dense oxide and lead oxide particles. This reduces the current but means the battery can be completely discharged and recharged many times.*

Creating a deep discharge battery requires a different electrode design than for an SLI battery. In these applications, the goals are a battery that can be fully discharged and that has a higher energy capacity than a car battery. While a car battery is a high-powered battery, here we seek batteries that can supply steady, long-term amounts of energy and be completely discharged without damage. Instead of being used to turn on a car, these batteries might be used to keep a car running for miles at a time, or they might collect energy from solar panels and then power a house until they are drained. Typically, in a car battery, the electrodes are spaced further apart and are made thicker to increase energy capacity. These batteries also have a space below for debris to gather so that when the lead sulfate forms on discharge it can slough off below the electrodes. This eliminates the problems that occur when an SLI battery is fully discharged.

These differences highlight the most acute design issue that lies at the heart of any battery: The trade-off between energy density and power density. Thin electrodes give a great deal of power (low resistance), but to make them thin you need more inactive material to support them, hence lower energy density. Thicker electrodes have more active electrode material per unit volume, but with that comes a higher resistance, so they have a lower power density but higher energy density. In a deep discharge battery, we trade weight and compactness for higher energy density to be able to recharge it after being completely drained. This creates a heavier, bulkier battery than a shallow charge SLI battery because the deep discharge battery has extra room at the bottom of the battery and thicker electrodes.

The First Lead-Acid Battery

Gaston Planté (1834-1889), a French physicist, built the first lead-acid battery that went beyond a laboratory curiosity. A child prodigy —he spoke fluent English, German, Spanish, and Italian and read Greek and Latin by the time he was 16—demonstrated great talent for drawing and music but chose to spend his life as a scientist. At age 20, he discovered fossils of a large species of a flightless bird that lived 55 million years ago, later named *Gastornis parisiensis* in his honor. But he loved constructing electrical apparatus. He had such a talent for building devices that Emperor Napoleon III and his wife invited him to demonstrate an electrical curiosity popular at the time: an induction coil that used long spirals of copper wire to create two-inch-long brilliant and astonishing sparks. His experiments bore practical fruit when, in 1860, he perfected his first lead-acid battery: It consisted of nine cells made by coils of lead separated by a sheet of flannel and covered by sulfuric acid and water. An unusual man, Planté lived only for his experiments; he never protected his inventions by patents and showed little interest in earning money from them. He declined to join the prestigious French Academy of Sciences, noting that it would take too much time to prepare for the election, time he'd rather use to work in his laboratory. He wrote in his journal of his meeting with Napoleon that his being "mixed up with royalty" did not impress him. Although declining worldly recognition, Planté was generous. He spent much of his fortune helping impoverished scientists. In his will, he left three valuable pieces of real estate to the Friends of Science Society.

Like many successful technologies, Planté's invention showed only the possibility of a practical battery. It needed to be refined and other inventions had to be created to make it become a permanent fixture in our technological firmament. The battery's real limitations lay in its electrodes. Planté used lead as both the positive and negative electrodes. This meant that charging the battery took days because one needed to convert the lead to lead oxide. French engineer Camille Faure (1880-1881) created the battery we have today by coating the electrodes with a thick layer of "red lead," which could be converted easily to lead oxide. This improved battery could be charged quickly. Yet even with this improvement, the battery could not easily be charged because it was invented before mechanical generators of electricity—the dynamo that still creates electricity in power plants. The lead-acid battery could only be charged with high-resistance batteries that were difficult to build and use. Lastly, the lead-acid battery needed a "killer app" use. Its first uses, like any new technology, were pure novelty: In 1881 it powered

> a three-wheeled car, in 1886 it propelled a submarine, and in 1899 it drove a cigar-shaped electric car at an astonishing 67 miles/hour. The lead-acid battery came to prominence, and apparently permanence, when the emerging telecommunications industry adopted it for powering telegraphs.

Why Does the Lead-Acid Battery Still Exist?

While the lead-acid battery may sound like a quaint nineteenth century invention, it is still a vital part of our lives today. Lead-acid batteries are a 30 billion dollar market, with three-quarters of them used for SLI, mostly for cars. In our high-tech age, it seems obvious that we should replace such a heavy and acid-filled power source, but there are two key reasons it still thrives.

The primary reason we keep this seemingly old-fashioned power source around is simple: It's cheap to build. Made of mostly lead and acid wrapped in plastic, its materials are cheap to manufacture: A lead-acid battery costs, for example, about 1/10 that of a lithium-ion battery per unit energy density. However, although it's cheap to make, almost no other battery designs have the high voltage and low resistance necessary to generate a large current and maintain a high power density. To produce the several hundred amps needed for ignition, the entire battery needs to have an internal resistance of less than 0.01 ohms. A few other batteries might work well for SLI but would not meet both these criteria. For example, some of the most advanced lithium ion batteries can have power densities exceeding that of a lead-acid battery, but lithium sells for about thirty times the price of lead per pound. A nickel-iron system would be nearly as cheap as the lead-acid system, but it has extremely high internal resistance and thus a power density of about half that of the lead-acid battery. More simply put: A nickel-iron battery cannot supply the large bursts of currents needed without being much heavier than a lead-acid battery.

But why is the lead acid battery able to supply currents that other batteries can't match? It isn't something in the design, or else we would see that in all other batteries as well. Instead, it is intrinsic to the lead-acid system, a natural property of the chemicals used. The answer is actually pretty simple: A lead oxide electrode has an extraordinarily high conductivity compared to the positive electrodes in most batteries. For example, lead oxide conducts electricity orders of magnitude greater than the positive nickel oxide hydroxide electrode used in a nickel-iron battery. This high conductivity is the same, of course, as low resistance, which is at the heart of a good SLI battery.

So, we still have the lead-acid battery because of what nature handed us: plentiful, cheap materials and an electrode material very high in conductivity. Once we arrange these materials correctly, we have a cheap battery with a high power density. No other materials found in nature meet these criteria to create an SLI battery, which means that with current technology it is extremely hard to overcome this barrier. The SLI battery is less than ideal, though, for any other use. Imagine lugging it around, for example, to power a mobile phone; for that we require an ultra lightweight, but high-energy battery. The material of choice for that battery is lithium.

How Lithium-Ion Batteries Work

Engineers focused on developing lithium batteries because of two unique properties of lithium. First, of all the metals, lithium has the largest energy difference between its pure metallic state and its ionic state (Li^+). Lithium has a strong tendency to donate its electrons, so it is never found as a pure metal in nature and reacts strongly with water. This also means that when used as the negative electrode in a battery, lithium will produce the largest voltage. The second reason lithium is used in smaller applications like cell phones and remote controls is that it is the least dense metal. Combining this property with lithium's ability to produce such a high voltage makes it the perfect material for small batteries. You might think this makes it ideal for use in cars. Regardless of whether it could produce the current burst needed like lead, a lithium-based battery would be too expensive.

The first batteries used pure lithium metal but presented safety problems because lithium metal reacts violently with water. For instance, if the battery were pierced, it would explode. To avoid the problems with pure lithium metal in a consumer battery, engineers created a way to store lithium ions without having bulk lithium metal. They use what is called an "intercalation" electrode. These are compounds—graphite for the negative electrode and CoO_2 for the positive electrode—with tunnels or channels that allow Li ions to be inserted and removed as shown in the next figure.

To create such a battery, a thin layer of powdered $LiCoO_2$ metal oxide is mounted on aluminum foil for the positive electrode; for the negative electrode, a thin layer of powdered graphite is mounted on copper foil. Just as in the lead-acid battery, the two electrodes are separated by a plastic film so they do not come into contact. A lithium-based battery cannot use water as the solute because of the aforementioned reactivity, so an organic solvent must be used. This makes the battery cost more

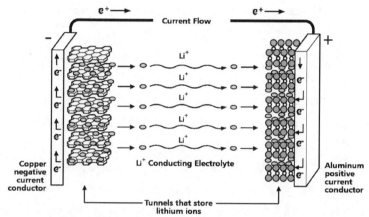

At the negative electrode—graphite—the lithium becomes embedded within the graphite structure as shown in the figure: $C_6 + xLi^+ + xe^- \rightarrow Li_xC_6$. At the positive electrode, lithium ions can leave the channels of the cobalt oxide that traps them: $LiCoO_2 \rightarrow Li_{1-x}CO_2 + xLi^+ + xe^-$. When discharging, the lithium atoms move from the tunnels in graphite to those in the cobalt oxide, and the reverse happens when charging. Because of the motion of the lithium ions, these batteries are often called "swing" or "rocking chair" batteries.

(the organic solvent is much more expensive than water) and also limits the choices because it must dissolve lithium salts. The plastic separator in a lithium-ion battery typically contains $LiPF_6$ dissolved in a mixture of organic solvents like ethylene carbonate (EC), ethyl methyl carbonate (EMC), or diethyl carbonate (DEC). These expensive organics are used because the lithium ion will react strongly in the water, yet we need something like water to dissolve the salts. This means big, highly polar organic molecules like the solvents EC, EMC, and DEC.

Why Lithium Laptop Batteries Explode

If not designed or used correctly, a lithium battery can explode. This is partly due to its high energy density but also because the materials used in it—lithium ions, cobalt, and organic solvents—can lead to a lethal cocktail. While an engineer might like to use other materials, these are the only ones that provide the necessary high voltages. In a lithium-ion battery, all the energy can be released at once by an internal short circuit. Just think of the damage caused if you short out an electrical outlet at home. This short circuit releases energy, causing uncontrolled flows of current, which produces heat that makes the temperature rise inside the battery. A lithium-ion battery can heat itself to over 700 ºC in a matter of minutes. (This is called *joule heating*: heat produced by a current flowing through a conductor.) These high temperatures supply the energy necessary to sustain a host of new reactions.

For instance, the lithium in both the positive and negative electrodes can react with the organic electrolyte to give off ethylene gas. Also, the lithium phosphate salt breaks down in the organic solvent, catalyzed by the cobalt in the positive electrode, to create PF_5 gas. All of these reactions are highly exothermic; that is, they give off heat as the reactions occur. This leads to what is known as thermal runaway inside the battery. The reactions begin to generate heat which, in some batteries, produces heating rates as high as 400 ºC/minute. This, in turn, increases the rate of the reactions, creating gases until an explosion occurs.

To prevent an explosion, lithium batteries have two key safety mechanisms. First, the separator is designed so that at 125 °C or so its pores close up, thus preventing the flow of ions and stopping the current flow. However, if this separator is pierced, this mechanism won't work. So lithium-ion batteries have sophisticated electronic controls that keep track of the voltage and shut the battery down if it reaches a dangerous voltage. Although they are manufactured with safety in mind, lithium batteries can also have design flaws that allow them to explode. In 2006, Sony had to recall almost six million laptop batteries used by nearly every major laptop manufacturer because the batteries were found to fail 1 out of 200,000 times. Although this might seem small, in battery manufacturing, safety incidents with correctly made lithium-ion batteries only number one in ten million. A manufacturing defect caused contamination of the batteries by metal particles, which pierced the separator, creating an electrical short circuit that released the battery's energy rapidly, resulting in an explosion. Small wonder the United States Department of Transportation classifies lithium-ion batteries as a hazardous material for shipping.

In Depth: Entropy

The First Two Laws of Thermodynamics
I. Energy can be neither created nor destroyed.
II. The entropy of the universe must always increase.

Now that we have discussed batteries, let's take a minute to think about thermodynamics. Now, the mention of the word thermodynamics may scare some of you, but we assure you it is a field with which you are all very familiar. For instance, inside of batteries, solid lead dissolves in a sulfuric acid/water solution and gives up an electron, and that is part of what allows us to get energy from batteries. But why does it work like that? What exactly is the "force" that makes it so lead dissolves in sulfuric acid, or salt dissolves in water, or makes anything happen really? Well, to explain that we need to share with you how to think about entropy and the second law of thermodynamics.

To begin, we should re-phrase this question in a very careful way; if you ever study thermodynamics, be aware that precise language must be used or you'll get the wrong result. Here's the proper question: "Why does lead spontaneously dissolve in sulfuric acid and give up an electron?" The key addition we've made is the word "spontaneous." What we're looking for is the direction of natural change. Intuitively, you're aware of spontaneous or natural change: Try to balance a cone on its point and it's likely to fall over; drop a tennis ball, and it will bounce until it decays to stillness; or set some ice on a counter, and it'll melt if you don't do anything else (keep that "anything else" in mind because we'll return to it in a moment). You would be surprised if suddenly, on its own, a cone stood up on its point, or a ball started bouncing, or ice went from a liquid to a solid, because these are just not spontaneous changes. So how, in a scientific way, do we characterize that direction of spontaneous or natural change?

Let's explore an answer that's wrong but common: "The direction of spontaneous change occurs—for example, the exchange of electrons in a discharging battery, or the melting of ice—because the total overall energy is being lowered. Things naturally seek a lower energy state. So, the direction of spontaneous change is toward lower energy." Now, after reading this go back and re-read the first law of thermodynamics above. These laws were articulated in the nineteenth century and have been proven time and again. A moment's reflection on the first law of thermodynamics should convince you that our first attempt at an explanation for spontaneous change is not correct.

The first law boldly states, "energy can be neither created nor destroyed." So, return to any of our examples—dissolving lead, bouncing balls, melting ice—and consider the implications of this first law. If energy is neither created nor destroyed, then even though the bouncing ball lost energy, which it surely did, everything else in the universe gained energy. The net difference was zero! No doubt you might be thinking, "Yes, but the ball's energy did decrease, and that's why it stopped bouncing." And we ask you: "Why did we choose to focus on the ball's energy and not that of everything outside of it?" Therein lies the key to understanding the second law and entropy at an intuitive and physical level.

When the ball bounces, it distributes its energy a little bit on each section of the floor in such a way that we don't notice the floor gaining energy. But what distribution of energy defines the direction of spontaneous change? The first law simply says this: If you add up all these little bits of energy, you'll find that energy was neither created nor destroyed. From the perspective of the first law alone, it's perfectly fine for the ball to start bouncing, it would just need to gather energy from the floor! That's why we need the second law.

The second law tells us that the direction of spontaneous change is toward the most chaotic distribution of energy in the universe. That is, energy will attempt to "even out" or be the same value everywhere. The ball, for example, has a great deal of potential energy when we hold it, but when we let it bounce, it dissipates its energy across the floor. So, to quantify this in a way that allows engineers to design things like air conditioners and engines, we need a measure of how much that energy has been dissipated. The degree to which energy is distributed is called entropy.

Often we will hear someone say that the second law means that things must become disordered. We'll hear someone just reel off "entropy must increase." Yet by observing the world around you, you can see that entropy can decrease! The creation of life, for example, would seem to violate this second law, as complex structures are created from individual cells (we don't actually violate the second law, however, as we get energy from food). As we learned with a bouncing ball, entropy is a measure of how disordered energy has become. Intuitively, this means that at its freezing point, liquid water will have a higher entropy than solid ice, and steam will have an even higher entropy than liquid water, because steam is so much more disordered than liquid water, which is even more disordered than ice. Indeed this is true: At the freezing point, the entropy of liquid water is 1.7 times that of ice, and water vapor has an entropy 4.6 times that of ice. So, have you ever made ice? Of course you have, which means that you have lowered the entropy of something! So, what's sloppy about characterizing the second law as "entropy must increase" is that you must include the phrase "of the universe." Let's explore this a bit more.

Recall that we carefully said above, "Put ice on a counter and it'll melt if you don't do anything else." You can, of course, keep ice from melting by doing "something else," namely, putting it in a freezer. This may seem to violate the second law, but of course it doesn't. To keep that freezer running, you must do that something "else," namely, generate electricity at a power station. If we look at the complete picture of ice + refrigerator + power station + everything else, then the entropy of the universe has increased. No wonder that sometimes the second law is stated as, "You cannot get something for nothing."

This applies to chemical reactions as well. As we covered in the previous chapter, chemical reactions occur at the electrodes of batteries, causing a transfer of electrons, which we use to power electronic devices. These and every other chemical reaction occur because they allow for the overall entropy of the universe to increase. Hence, entropy is said to define an "arrow of time" because if time is moving forward, then entropy must be increasing. Keep this in mind as you go through your day-to-day life and think about how everything you do increases the entropy of the universe just a little bit more. Isn't thermodynamics wonderful?

Aluminum
[al-oo-min-um] | Symbol: Al
Atomic number: 13 | Atomic weight: 26.981534 amu

Aluminum derives its name from the Latin alumen *meaning alum, a bitter-tasting salt of aluminum. A piece of pure aluminum shines with a bluish-white metallic luster and can be easily bent and shaped. Although it is the third most abundant element in the earth's crust behind oxygen and silicon, it is never found pure in nature. Locked inside bauxite ore, at first aluminum could only be extracted with great effort. It was so expensive that Napoleon considered it a precious metal and Europe's royalty had "silverware" created from it. Aluminum became ubiquitous when two chemists discovered how to use electrolysis to extract it from bauxite, thus dropping the price by a factor of nearly 100. With this innovation, aluminum moved from a status symbol for the rich to becoming commonplace, even disposable: Think of the everyday soda pop can or the wrapping for your leftover food. In this chapter, we discuss*

Hammack, Ryan & Ziech

an amazing property of aluminum: How engineers can electrochemically grow a diamond-like layer that can encapsulate the vivid colors seen in laptops and MP3 players.

Anodizing, or The Beauty of Corrosion

I LOVE APPLE'S BEAUTIFUL "unibody" design used in many of their laptops. Although each laptop is machined from a single piece of aluminum, they all have a diamond-like toughness with a polished and refined look. They use a similar look on their iPods, but with an embedded color; you can find iPods made of aluminum glistening in many colors. Although it looks like a painted coating, it is actually an integral layer grown into the aluminum. The trick to this amazing blend of beauty and utility can be learned from a pillar that dates from 400 AD.

What Is Unibody Design?

Two main methods exist to build a structure: a space frame or unibody construction. (By structure, we mean anything that supports weight, like a laptop body, an aircraft wing, or a car chassis.) In a space frame, an engineer uses separate struts and tension rods to build a lattice that supports weight (the "load" in engineering parlance). Think of a child's erector or Meccano set. Often this lattice is covered with metal sheet or cloth, but these elements bear no significant weight. Most familiar is the girders used to construct a building. In contrast, a unibody (also called a monocoque structure) supports the load completely in the panels that make up the structure. Inspect an aluminum beverage can: You'll see no lattice inside supporting it; instead, the can's body sustains the load completely. Nature also uses this type of construction, as in the lobster's shell.

> To many people, a unibody design conveys a sleek, modern look, especially in the case of Apple's products. Many assume monocoque to be superior to the space frame. Yet, as with any engineered object, a series of trade-offs determine what type of structure a designer uses. A space frame, for a structure with mostly compressive loads, is always lighter and thus cheaper than a unibody design. Small wonder then that rarely do large animals have exoskeletons, which would be unibodies. Most, instead, are vertebrates and thus space frames. It is only when significant torsion occurs, which nature always avoids in its designs, that a monocoque structure makes sense.
>
> Consider aircraft. Early planes used space frames covered with cloth. While a space frame resists compression and bending more efficiently than a unibody, the latter resists shear and torsion better. Airplane speeds increased, along with shear and torsion, until by the 1930s it made sense to "pay" the extra structural weight for a monocoque design. All modern jets use this type of design. Still, when weight becomes absolutely critical, the space frame wins out, as in modern hang gliders.

The Delhi Iron Pillar, a pillar made of iron nearly 1,700 years old, sits in the Courtyard of the Quwwat-ul-Islam mosque near the Qutub Minar tower in New Delhi, India. Weighing some 13,200 pounds, standing 24 feet high, and tapering slightly from a roughly 16-inch base to about 13 inches at the top, it's capped with an aesthetically pleasing finial. (In si units it's 6,000 kg, 7.4 meters high, tapering from 42 centimeters at the base to 34 cm at the top.) Even more technologically amazing than the outer skin of an Apple laptop, this pillar has stood for ages without decaying. Think of that for a moment: A gigantic piece of iron has been exposed for centuries to rain, sun, and wind, yet none of it has rusted. Any other piece of iron from that era has long since been corroded into useless bits scattered by the wind. The secret to this structure is not that it hasn't been corroded but that it has been corroded in a way that preserves it forever.

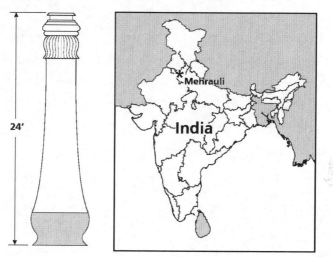

Although it looks simple to us today, this pillar is one of the great engineering achievements of antiquity. Until the nineteenth century it was the heaviest known piece of wrought iron in the world. Likely built in the late fourth or early fifth century AD, the pillar has stood tall for nearly 2,000 years. Its iron contains a high level of phosphorus, which creates an impenetrable protective film.

We think of corrosion as a force of destruction, but when used properly in a creative way by engineers, like the corroded aluminum body of an Apple device, it yields beauty and utility. Likely unknown to the ancient Indian metalworkers who built the Delhi Pillar, its iron contains a great deal of phosphorous. When the iron corrodes, the phosphorous combines with the iron to make a protective film that seals the surface, effectively keeping out moisture and other chemicals that would destroy the iron. The Delhi Pillar has also benefited from being in a part of the world where the repeated wetting and drying cycles have, over time, converted the surface film into nano-sized regions that further help to reduce rusting. Similarly, the surface of the aluminum used in Apple's laptops or iPods has a similar protective film that grows naturally, but that's because engineers use a series of electrochemical reactions to strengthen the film until it becomes extra thick and robust. Those protective layers in the phosphorous-rich iron of the pillar and the nearly diamond-like aluminum oxide come about in the same way as a more familiar destructive force, rust.

Why Do Metals Corrode?

Really the question of "Why do metals corrode?" isn't that interesting. It's better to ask, "Why don't they corrode more quickly?" After all, rarely do we find pure metals in nature; they usually combine with something. The exceptions are only a handful of noble metals: platinum, gold, silver, and palladium, as well as the more exotic ruthenium, rhodium, and osmium. Most metals, though, are highly reactive. Let us explain.

Place a bit of pure iron in water, and the very familiar reddish, flaky rust will begin to form. This red substance is an iron oxide, although other kinds (and colors) of "rust" exist. For instance, underwater iron will combine with chlorine to create green rust. These days, engineers typically replace the colloquial "rust" with the more general term "corrosion product." Corrosion occurs because electrons transferred from the metal are taken up elsewhere and the ions formed by this electron exchange either combine with ions of opposite charge or are removed. If these charged ions are not removed, an accumulation of charge occurs, which would inhibit further corrosion. Knowing that corrosion involves the movement of electrons partly explains why metals corrode easily; what better way to move an electron than with a metal? Now, let's look at corrosion with a specific example.

Expose pure iron to the atmosphere where water and oxygen are plentiful, and the iron will give up its electrons

$$Fe \rightarrow Fe^{2+} + 2e^-$$

and the acid will take them up

$$O_2 + 4e^- + 2H_2O \rightarrow 4OH^-$$

As we mentioned earlier, the ions formed need to be either swept away or recombined. Typically, the iron combines with the oxygen and hydroxides to eventually form all sorts of colorful iron oxides: FeO and Fe_2O_3, among others. For example,

$$Fe(OH)_2 \rightarrow FeO + H_2O$$

Oxidation-Reduction Reactions

The general name for these types of reactions is "oxidation-reduction," but they are commonly referred to as redox reactions for short. These names came about long before chemists truly understood atoms, electrons, and chemical bonds. Early scientists noted, for example, that when pure calcium is heated with oxygen, the following reaction occurs:

$$2Ca_{(s)} + O_{2\,(g)} \rightarrow 2CaO_{(s)}$$

Because the first redox reactions observed involved oxygen, chemists developed a way to describe the changes occurring based on oxygen. They used the term "reduced" to describe what happens to the oxygen in the reaction above: It has been "reduced" to its elemental form. Thus, they describe its oxidation number, using a simple scheme described below, as changing from 0 to -2. Next they describe the calcium as being oxidized. That is, oxygen was added to it. Here the oxidation number of calcium has changed from 0 to 2. Assigning the oxidation number—more correctly called the oxidation state today—involves a few simple rules.

Any pure element (even if it forms diatomic molecules like oxygen, O_2) has an oxidation number of zero; for example, oxygen and calcium in the reaction above.

For monatomic ions consisting of only one atom, the oxidation number is the same as the charge of the ion; for example, the oxidation number of Li^+ is +1.

The sum of oxidation numbers of each atom in a molecule or complex ion equals the charge of the molecule or ion. This means that the oxidation number of an element in the molecule or ion is calculated from the oxidation number of the other elements. For example, in CaO, the total charge of the ion is 0, and each oxygen is assumed to have its usual oxidation state of -2. This means the oxidation number of calcium is +2.

Often we think about such redox reactions as involving a change in electrons, but many reactions do not involve changes in the formal sense. For example, when carbon dioxide reacts with hydrogen:

$$CO_{2(g)} + H_{2(g)} \rightarrow CO_{(g)} + H_2O_{(g)}$$

There is no change in the valence electrons, but the oxidation number of the hydrogen increased.

Redox reactions abound in technology, such as explosives consisting of strong oxidizing and reducing sections within the same molecule.

Household bleaches like Clorox have chloride-based acids that oxidize the color-bearing substance in stains. In a car, and likely in the power plant that creates the electricity that runs your home, a hydrocarbon with multiple bonds "burns" with oxygen, thus liberating energy. Here the gasoline acts as a reducing agent and the oxygen as an oxidizing agent:

$$2C_8H_{18} + 25O_2 \rightarrow 16CO_2 + 18H_2O$$

And in the car, the lead-acid battery starts it with a jolt of electricity from the redox reaction of lead and lead oxide.

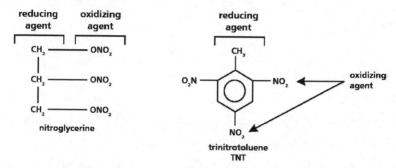

Typically an explosive contains both oxidizing and reducing sections within the same molecule. Shown on the left is nitroglycerin used in dynamite, the first practical explosive ever produced that was stronger than black powder. On the right is trinitrotoluene, or TNT, one of the most commonly used explosives because of its insensitivity to shock and friction.

How to Control Corrosion

Since corrosion occurs naturally, engineers have long since learned techniques to deal with the problem. The three ways engineers handle corrosion include controlling it to minimize damage, stopping it to completely halt damage, or using it to make a surface that is more corrosion-resistant than before.

The first strategy is to adsorb a monomolecular (super-thin) layer of a chemical compound called an inhibitor onto the metal surface to keep oxygen and moisture from acting on the surface.

Typically, the anti-freeze used in car radiators has sodium benzoate added to it for this purpose. The exact mechanism of how this layer prohibits corrosion isn't completely understood, but it's believed that the sodium benzoate enhances the growth of an oxide film that acts as a barrier. To put it in terms of erosion, a force we are more familiar with, this strategy would be analogous to laying a tarp on a field to prevent rain from washing away soil.

A second strategy is to redirect the flow of electrons to prevent corrosion. Often engineers do this in the most obvious way—by using a battery! The giant offshore platforms that extract oil from the ocean floor use this. The underwater structure supporting the platform is a large steel column. Although pretty resistant to corrosion, the harsh environment of sea salt surrounding it for years inevitably takes its toll. To avoid catastrophic corrosion, engineers connect the steel structure to an electrode and use a battery to change the flow of electrons. This method is analogous to putting sewage pumps in a field, to pump out rainwater to prevent it from taking away the soil.

A third and final strategy to prevent corrosion is with an impenetrable coating. This, of course, returns us to the Delhi Pillar and Apple's aluminum case. These coatings differ from an inhibitor; rather than temporarily coating the surface, these liquid coats are permanent. One could simply paint the surface, which is often done in cheaply manufactured objects, but a better idea is to grow the protective layer from the material itself. If done this way, the protective layer is attached by strong chemical bonds within the material and, in the best cases, can regrow if the top layer is scraped away. Such a process explains why the Delhi

Pillar stays intact, why our buildings don't fall down, and the beautiful durability of Apple's laptops. This strategy is analogous to planting bushes in a field to absorb rainwater. The rain makes the bushes bigger and stronger, which in turn makes it more difficult for rain to wash away soil.

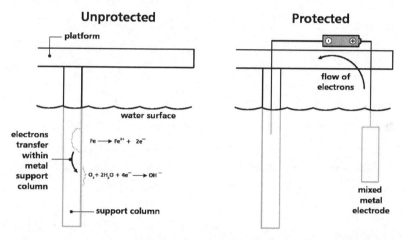

The structure on the left *is unprotected. Without cathodic protection, the iron in the steel pillar can corrode. In salt water, the elemental iron loses two electrons to become an ion, which dissolves in the water. The two electrons produced are conducted through the pillar, where the dissolved oxygen and water react to form hydroxide. These can form $Fe(OH)_2$ and the familiar oxides that appear when iron rusts. The structure on the* right *is protected. By supplying electrons to the pillar, we can prevent the iron from dissolving. Typically an offshore rig will have a large metal oxide electrode next to the pillar. A battery drives electrons from the new electrode to the pillar.*

> ### Why Our Buildings Don't Fall Down
> ### (Or, Why Stainless Steel Doesn't Rust)
>
> Although mainly composed of iron that typically rusts rather easily, stainless steel corrodes very, very slowly. Called the "Miracle Metal" and the "crowning achievement of metallurgy," stainless steel truly makes our modern world possible: Buildings, ships, cars, and likely even your kitchen sink are made of the stuff. Reflect for a moment on what a wonder it is: Imagine being an alchemist in the Middle Ages and being told that a material exists that looks like platinum, resists chemical attack like gold, and can support a quarter million pounds per square inch. You'd think the inventor was the greatest miracle worker of all time! The secret element in stainless steel is chromium.
>
> Stainless steel contains, by definition, at least 10% chromium by weight. Chromium reacts with water and oxygen to form a thin, tightly-bound stable protective film (similar to the one found on the Delhi Pillar and Apple's computers) that covers aluminum. Since the chromium is mixed through the iron, and not just on the surface, the protective layer will grow anew if the surface is scratched or if the piece of steel is cut to be shaped into something else.

Anodizing Aluminum

We typically think of aluminum as inert and unlikely to corrode or rust (think of aluminum foil or a bike frame), yet from a chemical viewpoint, aluminum is highly reactive. Sodium hydroxide dissolved in water will attack aluminum vigorously; calcium hydroxide gives off hydrogen almost immediately on contact with the metal. Hydrochloric and other halogen acids attack aggressively; when exposed to the chloride ions in sea water, aluminum will corrode within a few years. Yet we find it an inert metal in our technological world because engineers control the corrosion in a way that it forms a substantial protective layer.

Left exposed to the air, oxygen will attack aluminum's surface. The resulting aluminum oxide film keeps the aluminum from corroding further because it seals completely, at the atomic level, with a barrier film that doesn't conduct electricity. Recall that for corrosion to occur, electrons must move. The naturally-formed film is very thin and can be removed easily by scratching the surface. To create industrial-strength coatings engineers enhance this film electrochemically.

To do this, they make aluminum a negative electrode in an electrolytic cell. Current flows so that it drives the aluminum to a negative potential by using a second, largely inert, metal as an electrode. This current causes an aluminum oxide layer to form. At first it forms the thin "barrier" layer similar to what forms naturally. But then as the electrolysis process continues, the

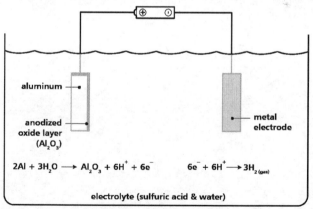

Anodizing occurs in an electrochemical cell. The piece to be anodized is used as the positive electrode. As electrons are removed from the positive electrode, a layer of Al_2O_3 grows into the aluminum part being anodized.

current "pushes" this barrier down into the aluminum, converting the aluminum above and growing a very porous oxide layer. This new layer is made from the aluminum: It isn't a layer being put on top but instead grows into it in a manner that consumes the aluminum. This is one of the reasons it's so effective at keeping aluminum from corroding. These pores give the aluminum a unique characteristic important for a consumer device: the ability to be colored or "decorated" to use the engineering term. The pores formed on the surface have a honeycomb pattern.

The center hollow section is typically 300 or 400 angstroms in diameter and perhaps 1,000 angstroms or so deep. (100 angstroms is a millionth of a centimeter.) Inside these layers, one can place dye of any color. Once the pores are filled, engineers

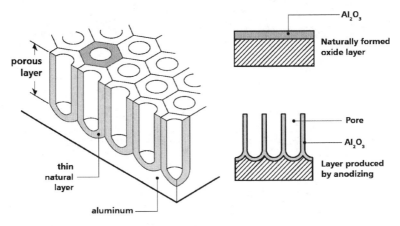

A piece of aluminum has a thin protective layer of Al_2O_3 that forms naturally on it. (See upper right-hand corner.) When anodizing takes place, this layer becomes thicker, then eventually forms a porous layer with hexagonal pores. Dye can be sealed into these pores.

seal the layer by boiling the aluminum in hot water. This closes the pores, locking the color in forever; you cannot scrape the color off without removing the aluminum, as it is a thick, tough oxide coating. Without the sealing, only the thin barrier lying adjacent to the metal would exist. Just like the natural barrier, it isn't thick enough to provide robust protection, but the sealed porous layer creates an extremely tough surface with or without dye. Its toughness likely comes from the oxide's relationship to tough gemstones. Sapphire is Al_2O_3—an aluminum oxide—with trace amounts of iron and titanium to give it a blue color; the same Al_2O_3 structure with chromium that absorbs yellow-green is also the basis of ruby. Both materials are very hard: 9.0 on the Mohs scale, which ranges from 1 for soft materials such as talc to 10 for diamond. Apple typically uses a thinner oxide layer resulting from what's called soft anodizing. This isn't as durable as the thick layers used for industrial products.

Anodizing Titanium

Titanium, a metal with many properties similar to aluminum, can also be anodized with beautiful results. Unlike aluminum, the thickness of titanium's oxide layer rather than dye determines its color. In titanium, a much thinner, transparent oxide layer forms than on aluminum, which gives titanium a wide array of colors due to thin-layer interference.

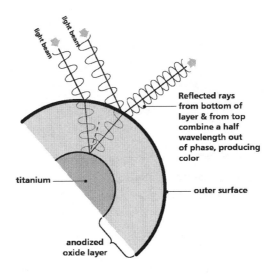

The oxide layers in titanium appear colored because of thin-layer interference. It's the same phenomenon that causes the iridescence in opals, oil slicks, soap bubbles, peacock feathers, and rainbow trout. When light rays strike the titanium surface, some pass through the transparent film created by anodizing and reflect off the surface, and others reflect from the outer edge of the film. If these two rays of a particular color combine when they are a half wavelength out of phase—where the crest of one meets the troughs of the other wave—the color observed will be white minus that color, thus appearing as its compliment. For example, the red in rainbow trout comes from the destructive interference of cyan, whose compliment is, of course, red. This means that the color of anodized titanium depends on the thickness of the layer.

Primer: Waves

THROUGHOUT THIS BOOK, we often mention waves in one way or another. Here we will cover a few basic principles of waves so that we can use waves to further our discussion of the electromagnetic spectrum.

Basic Definitions

Start by imagining a buoy in a shallow part of the ocean. It is, perhaps, 10 meters or so below the surface and can move up and down on a pole that keeps it from moving left or right. As you would expect, it will bob up and down with the motion of the ocean. If we were to measure its position on the pole over a long period of time, we would see the buoy rise to a maximum height, then fall to a minimum height, and repeat this motion again and again.

We can draw a line between the two extremes that neatly divides the buoy's position: Half the time it is above this line, half the time below. The deviations of the buoy from this level are called its amplitude. We call the time it takes for the buoy to start at its maximum positive amplitude, pass through the maximum negative amplitude, and then return to the maximum positive amplitude the period T; that is, the time between two "crests." In the drawing below, this happens in two-tenths of a second. Often, instead of talking about the time it takes for these two crests to appear, we'll talk about how many crests we would observe in a second. This is called the frequency, f, and is related to the period T by $f = 1/T$. For the buoy, the frequency is 5 crests/second.

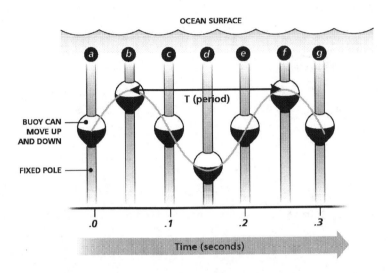

The motion of a buoy with time in a shallow part of the ocean. The buoy is on a rigid pole so that it can move up and down but not from side to side. The figure shows the same buoy at seven different times – a, b, c, d, e, f, and g.

Next, imagine a series of these buoys fixed along the ocean bottom. If we take a snapshot at various times, we can find the wavelength of the wave. We call the distance from one crest to the next in any snapshot the wavelength, usually designated by the Greek letter λ.

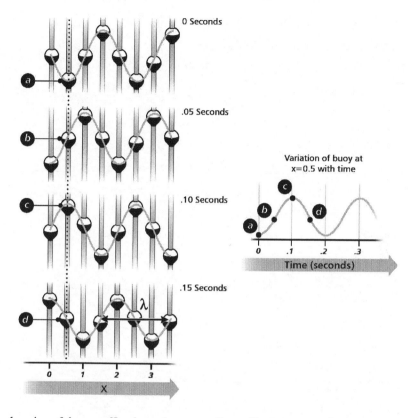

A series of buoys affixed to the ocean floor. Shown top to bottom are snapshots of the buoys at different times. Note that here the horizontal axis is the distance between buoys, where in the previous drawing it was time. You can construct the graph shown in that figure by looking at the motion of one of these buoys with time. For example, the positions of the buoy at 0.5 seconds—labeled a, b, c, and d—are shown on the graph on the right.

The wavelength and the period (or frequency) are related. The period is the time it takes a wave to travel one wavelength. Unsurprisingly, then, these are related by the velocity, v, of the wave: $\lambda = vT$. In the ocean, at a depth of 10 meters with a 1 m wavelength, the velocity of the wave will be 1.25 m/sec.

A wave, of course, transports energy. You can see this by the motion of the buoy: The wave's energy slides the buoy up and down the pole. More profoundly, think of how the ocean's waves erode a beach.

Interaction of Waves

We've talked above about a single traveling wave so that we could clearly define its properties, but the most important aspects of waves occur when they interact. Two or more waves can traverse the same section of space independently. Since they are independent, we can just add their amplitudes at the points in space or time where they overlap.

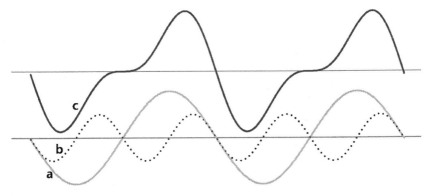

*The superposition (combination) of two waves (*a *and* b*) to create a new wave (*c*). To construct the new wave simply add the amplitudes of the individual waves.*

Technologically, the most important interaction of two waves is to form a standing wave; it's what heats the food in a microwave oven and what allows for the amplification of light in the cavity of a laser. It's best seen in a one-dimensional wave like we've discussed above. Picture one wave traveling along a line from the left to the right side of the page and one wave traveling along the same line but in the opposite direction. To keep things simple, assume the waves have the same wavelength. Now, if they start from different distances from the left and right, they will be out of phase and the waves will cancel each other out.

If, though, they started from spots equidistant from the center (and we'll explain in a moment how that happens), they will meet in phase, and thus the crests and troughs of the waves will match up. The result will be a wave that has points that are always zero, called nodes, and the resulting combined wave will oscillate between these points. If we imagine a particle sitting on a section of the wave, it would rise and fall but not move left to right.

In engineered devices like the microwave and laser, a standing wave usually results from a wave being reflected. For example, this is done in a laser by the mirrored ends of the cavity and in a microwave oven by the metal walls. Specifically, reflection allows only waves with an integral number of half wavelengths to form standing waves. Waves of other wavelengths simply die out as the waves with allowed wavelengths grow larger and larger as their crests reinforce each other.

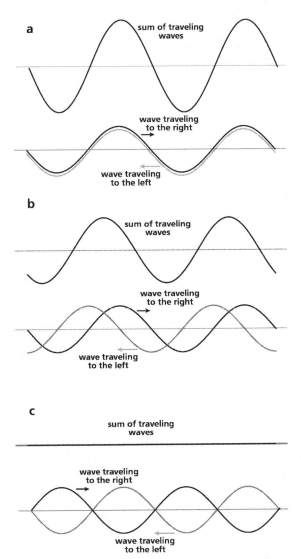

The bottom of (a) shows waves traveling from the right and left, the top shows their sum. (b) Shows the two waves shifted relative to each other by 90°. (c) Shows the two waves shifted 180° out of phase with each other. They cancel each other so their sum creates a "wave" of zero amplitude.

Eight Amazing Engineering Stories

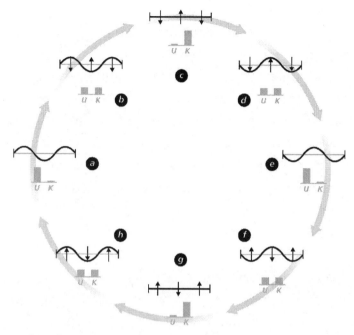

Shown above is a complete oscillation of a standing wave. The small graphs show how the total energy of the wave is distributed between potential energy (U) and kinetic energy (K). A standing wave does not appear to move; instead, the traveling waves that comprise it superimpose to create a wave with fixed nodes—spots where the wave's amplitude is always zero— and anti-nodes, which smoothly change from the maximum positive amplitude to the maximum negative amplitude. A particle sitting atop the anti-nodes would be bounced up and down. It would come to a full stop as the anti-node reaches its most extreme points—the maximum positive or negative amplitude. At that position, the particle would have its energy stored completely as potential energy. As the anti-node changes, the particle would have its maximum velocity just as the wave passed the horizontal axis. Here its energy would be completely kinetic. Compare this figure to behavior of an LC-circuit as described in the chapter on microwave ovens.

Electromagnetic Waves

In this short primer, we've only shown a transverse traveling wave. The wave is transverse because the wave motion is perpendicular to the direction of travel. This type of wave motion is easier to visualize than longitudinal waves, where the motion is in the direction the wave travels. You can make both types with a Slinky: Fix it at one end and shake it side to side, and you have a transverse wave; in contrast, pulsing the fixed end in the direction of travel creates a longitudinal compression wave where the curls of the Slinky push against each other in order to facilitate energy transfer.

Waves come in many shapes and sizes, but usually engineers talk about sinusoidal waves, as we've shown above, that vary smoothly with time and distance as they pass by us. Any wave can be composed of the sum of many sinusoidal waves; thus an understanding of simple sinusoidal waves is all that's required to understand any wave.

Waves come in many different forms; for instance, you produce waves whenever you disturb water, pluck a string, or even make a sound! However, the waves we focus on in this book are electromagnetic waves. Electromagnetic waves differ from mechanical waves—for example, waves in water—in two key ways. First, they need no physical medium to travel through, so they can propagate through a vacuum. Second, they always consist of both an electric wave and a magnetic wave together.

An electromagnetic wave appears when either a magnetic field or an electric field oscillates. Regardless of how it's generated, once the wave leaves its source, it is a combination of electric

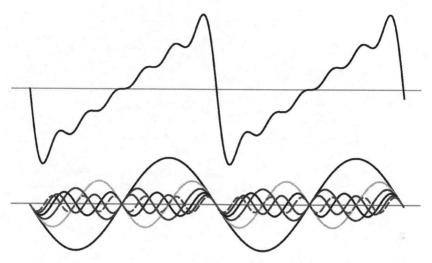

Any wave can be created by adding together an infinite number of sinusoidal waves. Here the five waves shown individually at the bottom of the figure combine to create the waveform at the top.

and magnetic fields in a well-defined relationship: In a propagating electromagnetic wave, one doesn't exist without the other. Electromagnetic waves of different frequencies make up the electromagnetic spectrum, which is the full spectrum of waves, categorized from smallest wavelength to highest. Although an electromagnetic wave may sound like a complex thing, we are all familiar with one electromagnetic wave: visible light!

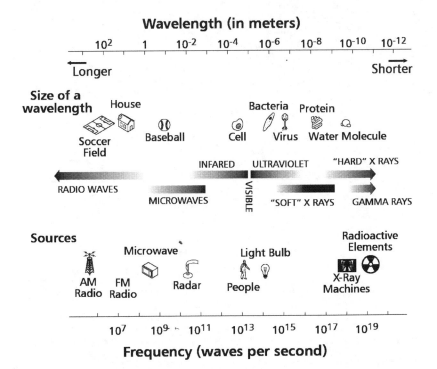

The electromagnetic spectrum contains all possible frequencies of electromagnetic radiation. It extends from low frequencies used for radio communication to gamma radiation at the short wavelength (high frequency) end; thus it spans wavelengths from thousands of kilometers to a fraction of the size of an atom.

Tungsten, Thorium, & Copper
[tung-sten] | Symbol: W
   Atomic number: 74 | Atomic weight: 183.85 amu
[thor-i-uhm] | Symbol: Th
   Atomic number: 90 | Atomic weight: 232.0381 amu
[kop-er] | Symbol: Cu
   Atomic number: 29 | Atomic weight: 63.546 amu

Tungsten began life as a nuisance: German miners nicknamed it wolfram *(wolf-dirt) because it interfered with the smelting of the tin they desired. The residue of that name is still evident in tungsten's chemical symbol "W". Later in life, chemists rechristened it* tung sten, *Swedish for heavy stone, an appropriate name for a material that rose to prominence as an element of war. Add tungsten to iron, and one makes an amazingly strong steel that was first used in creating cannons and guns for World War I. In fact, today's armament makers still use tungsten to make armor-piercing bullets. Aside from iron, tungsten is the metal most used industrially because of a single distinguishing characteristic: No other metal has a higher melting point, an astonishing*

3,422 °C. Thus, it can be heated to burn a bright white; small wonder it forms the filament of an incandescent light bulb. This same property makes it ideal for the vacuum tube powering a microwave oven.

However, tungsten alone doesn't produce enough electrons for its use in the microwave, so it's impregnated with thorium. Named for the Scandinavian god of thunder, Thor, this silvery-white metal has a low melting point but burns brightly since it is a good emitter of electrons. In the microwave oven's vacuum tube, thorium is combined with tungsten to get the best of both metals: a high melting point filament rich with electrons. The heated tungsten-thorium filament creates great heat, so it requires a material that can both conduct electricity and heat well. For this purpose, nearly nothing surpasses copper.

Copper is the most versatile metal on earth because it uniquely unites five properties—other metals have some, but none have them all—to make it perhaps the most important metal. It doesn't react with water, can be bent to any shape, plays friendly with other metals to form alloys, and conducts heat and electricity well. One would think the name of this powerful metal would come from some *marvelous Greek or Roman god, but instead it was named for the island of Cyprus (*cyprium *in Latin) where it was first found.*

How a Microwave Oven Works

PICTURE A VACUUM TUBE. Likely, you are thinking of an old-fashioned radio. Perhaps you imagine a huge wooden cabinet from the 1930s, filled with glowing glass tubes, with a vision of the inevitable replacement of these tubes by tiny transistors and microchips. Yet it's too soon to relegate vacuum tubes to the museum. Nearly every American home uses one every day in a microwave oven.

> **What is a Vacuum Tube?**
>
> Before the age of the solid-state transistor and the microchip, electronic devices used vacuum tubes, also called thermionic valves. In any electronic circuit, an engineer wants to control the flow of current to do something; for example, to amplify a signal or create an electromagnetic signal. (Contrast this with an electrical circuit, where current simply flows through a wire.) Vacuum tubes gave rise to the first wave of electronic technology, especially radio and television broadcasting. Today, of course, solid-state devices like transistors have largely replaced vacuum tubes. The solid-state devices last longer and are smaller, more energy efficient, and cheaper to make. An exception, as noted in this chapter, is when used in applications where greater power must be generated.

The vacuum tube, called a magnetron, produces electromagnetic radiation that heats food. Although the solid state revolution took over and miniaturized most electronic components, the microchip cannot easily replace tubes for producing power (recall that power is the rate at which energy is

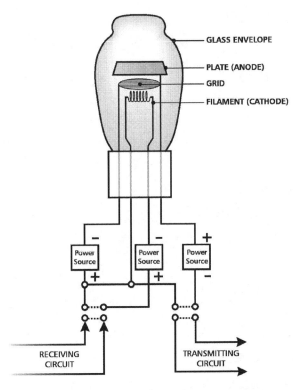

In a vacuum tube, an external power source heats a metal cathode so that electrons "boil" off. (This is what gives vacuum tubes their glow and also causes older electronics to become warm.) The positively-charged anode, called the plate, captures these electrons. Between them is a wire mesh called a grid. By applying a positive or negative potential to this, relative to the cathode, we can control the flow of electrons. A glass enclosure maintains a vacuum around these elements; without a vacuum, the electrons would not reach the anode. As shown above, it is wired to be an amplifier. The varying potential on the grid (the signal) is amplified by the electrons flowing from cathode to anode.

transferred). The signals transmitted from satellites require vacuum tubes, otherwise they would be so weak they could not reach the earth. The broadcast of AM and FM radio uses vacuum tubes. Radar used for weather forecasting or military surveillance must use vacuum tubes, as do the navigation systems used by commercial aircraft. All these applications require a great deal of power, so it's no surprise that the microwave oven, which needs power to heat food, also uses a vacuum tube.

As the radiation reflects back and forth from the metal walls of the oven, a standing wave will form inside the cavity. This standing wave causes hot and cold spots to form inside the oven, as the waves constructively and destructively interfere with each other, resulting in spots that have greater energy than others. Although the three-dimensional pattern of waves is difficult to predict, the principle can be seen by looking at the waves in a single dimension.

The wavelength of the microwaves used in a kitchen is 12.2 cm. This means that in a cavity 38.1 cm long (a typical size for a microwave oven), three peaks and three valleys—called antinodes—of the wave will fit inside. The peaks and valleys represent the greatest energy of the wave, while at the nodes where it crosses the x-axis there is zero energy. In our simple one-dimensional model, these occur at half-wavelengths—6.1 cm for a microwave oven. Nodes correspond to "cold" spots; that is, regions where there is no heating. You can see this in your own microwave oven. Remove the turntable and insert a glass tray covered with a thin layer of mozzarella cheese and zap it for fifteen seconds. When you remove it, you'll see that some of the cheese has melted but some of it has not. This shows the hot and cold spots corresponding to the anti-nodes and nodes in the standing wave.

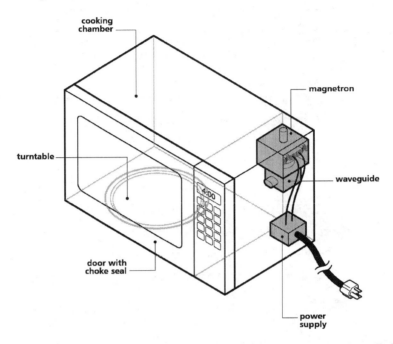

A microwave oven contains three main components: 1. a vacuum tube, called a magnetron, that generates high-frequency radio waves; 2. a waveguide that directs that radiation to the middle of the chamber, where the food is placed; and 3. a chamber to hold the food and contain the microwaves. The chamber has metal sides that the microwave radiation cannot penetrate.

Eight Amazing Engineering Stories

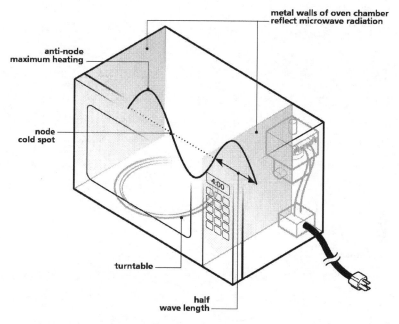

Only standing waves can exist inside a microwave oven, that is, waves whose half-wavelength is an integral number of lengths of the cavity. For the wave shown, three half-wavelengths fit inside the oven's chamber. The metal walls reflect the wave; at the walls, the amplitude of the wave is zero.

How Microwaves Heat Food

Many myths exist about how microwaves heat food, but in principle it's no different than any type of heating: At a molecular level, heat is a transfer of energy that results in increased motion of the molecules in a substance. Since we aren't quantum-sized, we observe this increase in motion as a rise in temperature. In a traditional oven or stove, we heat food by thermal contact: By the flow of hot air across it (convection), by setting a pot on a hot burner (conduction), or within the oven by infrared radiation. All of these methods heat the edges of the food and then cook the interior through conduction within the food. This is why, in a conventional oven, the outside of food cooks faster than the inside.

Two key differences characterize heating with microwaves. First, cooking occurs by volumetric heating, which means that instead of the outside being heated first, all of the food is cooked at once. In reality, this principle only holds true for perfectly symmetrical food of uniform composition, so most of the food you cook in your microwave won't be perfectly heated throughout. Second, the mechanism of heating is dielectric, which is explained in the caption below. To implement this type of heating, a household microwave oven uses electromagnetic waves that oscillate 4.5 billion times per second. These waves cause the water in the food to rock back and forth, and molecular friction resulting from this motion creates heat.

Microwave radiation rocks water molecules back and forth because the molecules are polar, meaning that their charge distribution isn't uniform. For example, a water molecule as a whole is electrically neutral, but this charge isn't distributed evenly across it. The electrons associated with the hydrogen shift toward the oxygen because of the eight positive charges in the nucleus. This leaves the oxygen end of the water molecule a bit negative and the hydrogen end positive. This uneven charge distribution, called a dipole, can be rotated by the microwave radiation. An electric field applies torque to a polar molecule, thus causing the rotation. Picture a water molecule between two plates of opposite charge; the negative plate will attract the positive end of the molecule, and the positive plate will attract the negative end, causing the molecule to rotate. In a fluid, a molecule will resist being rotated because it feels viscous drag from the other molecules around it. This drag causes friction, which generates heat.

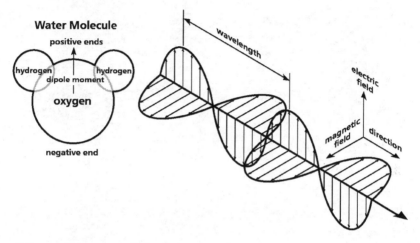

Microwave ovens use dielectric heating.

Does a Microwave Really Heat from the Inside Out?

In principle, microwave radiation will heat every part of a substance equally and simultaneously. This volumetric heating means that, in theory, a microwave oven doesn't heat from the inside out but rather heats everything at once. In practice, the situation is much more complex because no piece of food is uniform. There may be different water content within the food, or the freezing may be uneven. Also, the microwave radiation doesn't reach every section of the food. The microwave radiation may be attenuated or absorbed by the food's outer layers.

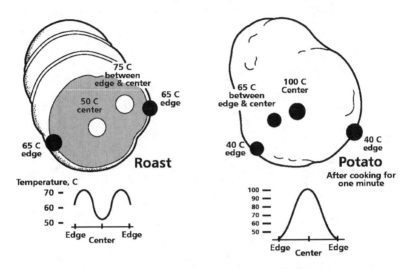

This figure shows temperature distributions inside food cooked in a microwave oven. Note that the heating throughout is not uniform because the food attenuates the radiation and because conduction occurs within non-homogeneous food.

For example, a slab of beef will be cooler on the inside than the outside, while a potato might be the opposite. This lack of uniformity in food is most noticeable when using the microwave to thaw food. If you use the cooking mode in a microwave oven when thawing food, you'll cook part of it while some of it remains frozen, because some sections of the food will always be unfrozen. This is because some sections contain higher concentrations of sugar and salt, which lowers the freezing point of water (This is the same phenomenon behind adding salt to an icy road.). These sections will absorb heat quickly and reach the boiling point of water before other areas thaw. This runaway heating results in partly cooked, partly raw food. To try to get the heating as even as possible, engineers build a defrost mode into the microwave oven. The only way of varying the amount of microwave energy entering the oven is to switch the magnetron on and off. Typically, kitchen microwave ovens cycle on and off every 15 to 30 seconds. To get a 1,000 watt magnetron with a cycle time of 30 seconds to halve its power (500 watts) for a defrost mode, the oven must turn the magnetron on for 15 seconds, then off for 15 seconds. This slower heating allows the quickly thawing sections to heat frozen sections via conduction so that all sections of the food are at roughly the same degree of thawing.

The Magnetron: The Powerhouse of a Microwave Oven

For a microwave oven, engineers need a device that produces electromagnetic radiation of 2.45 GHz (a wavelength of about 12.2 cm) with a great deal of power. The real engineering in the microwave oven lies in creating the magnetron that generates high-powered radio frequency (RF) waves. The magnetron is truly an amazing and revolutionary device; to understand how it produces microwaves, we need first to look at how any electromagnetic wave is generated. (For a short primer on waves, see the previous chapter.)

How to Create Electromagnetic Waves

The best place to start our understanding of electromagnetic waves is to examine two experiments and a groundbreaking theory from the nineteenth century.

In 1831, Michael Faraday did an experiment that changed the world. In essence, he passed a magnet through a loop of wire connected to an ammeter and noticed that the movement of the magnet caused a current flow. The French physicist Andre

Ampere showed, at about the same time, a complementary aspect.

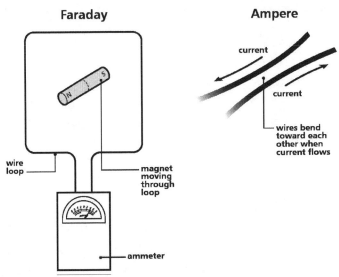

Ampere observed that two parallel wires with current flowing in opposite directions attracted each other; this meant there was a magnetic force between the wires. These two experiments showed that a change in magnetic field induces a current, and in turn the flow of a current generates a magnetic field.

Why a Microwave Oven Heats with 2,450 MHz Radiation

In principle, many different energies of radiation could be used to heat in a microwave oven. The reason we use 2,450 megahertz radiation in our ovens is a mix of scientific, social, economic, and political reasons—something true of almost any engineered object. In fact, a hallmark of something engineered is that it fits into our world—social, political, economic, scientific—unlike a scientific fact, which is either true or false. The only way to answer why a mass manufactured object operates like it does is to trace its history.

In the aftermath of World War II, all sorts of devices appeared that create electromagnetic radio—television, communication links, two-way radio, and radar. In the United States, the Federal Communications Commission regulated what devices could use what part of the electromagnetic spectrum. For microwave heating, which falls under the category of ISM applications (industrial, scientific, and medical), the FCC allowed two frequencies: the S-band (2,450 MHz) and the L-band/UHF (915 MHz). Each had a major corporate proponent: Raytheon, which made radar during the war, argued for 2,450 MHz, while General Electric wanted 915 MHz. Although Raytheon had made microwave ovens since the late 1940s, the sales of such ovens sputtered. In the 1960s, it wasn't clear who would make the first truly successful microwave oven. General Electric's ovens in the 1960s used 915 MHz radiation, while Raytheon stuck with 2,450 MHz. Which is better? It depends on what you are trying to do. Raytheon argued that 2,450 worked better for cooking small loads—like a hot dog—while GE argued that for large load (like a roast) penetration of 915 MHz was better and that it created less thermal runaway when defrosting. 915 MHz radiation penetrates food five times deeper than 2,450 MHz radiation, which means it can cook a large piece of food with minimal burning on the outside. But,

> as one Raytheon engineer said, 915 MHz radiation "won't cook bacon worth a damn." Also, the wavelength of 915 MHz radiation is nearly three times longer than for 2,450 MHz (32.8 cm versus 12.2 cm). This means that a typical 915 MHz oven would be larger than a 2,450 MHz oven; a 915 MHz oven might have to be four feet wide to be sure even heating occurred.
>
> In a way, social changes dictated the choice between these alternatives. With the rise in the 1970s of households where both parents worked, the microwave oven became a "re-heater" of small amounts of food, and thus a home kitchen appliance, rather than a large oven used commercially in a restaurant. These uses, then, made the 2,450 MHz radiation the standard for microwave ovens.

These two experimental observations were unified by Scottish physicist James Clerk Maxwell in one of the greatest and most technologically important scientific theories of all time. Maxwell gave a unified description of how electric and magnetic phenomena unite to form electromagnetic waves that can travel through space, even through a vacuum. Most of our understanding of electricity and magnetism comes directly from Maxwell's work.

In showing a continuum of electromagnetic radiation, Maxwell revealed something truly revolutionary: Electromagnetic waves could travel without a medium; that is, they could travel in a vacuum. Prior to his work, it seemed that a wave needed a medium such as a gas or a liquid to propagate in. For example, the waves in an ocean require the medium of water, otherwise that type of mechanical wave has no meaning. Maxwell's work opened up the notion that electromagnetic waves could fly through space, which led to the use of electromagnetic waves for radios, mobile phones, television, and so on. Further, Maxwell's laws reveal to engineers exactly how to produce and detect electromagnetic waves.

An electromagnetic wave appears when either a magnetic field or an electric field oscillates. Regardless of how it's generated, once the wave leaves its source, it is a combination of electric and magnetic fields in a well-defined relationship: In a propagating electromagnetic wave, one doesn't exist without the other. So, to create radio frequency (RF) radiation, we need to get an electric or magnetic field to oscillate at 3,000 times per second or greater. In principle, you could do this with a wire and a battery. Hook the battery up to the wire, let the current flow, and then reverse the flow by exchanging the wires on the battery's terminal. The problem is that radio waves oscillate at least 3,000 times or so per second; no human hand could move that fast! As you might expect, engineers design devices to make this switching happen without human intervention. For an engineer, the key in any device is to produce electromagnetic waves of the right wavelength or frequency.

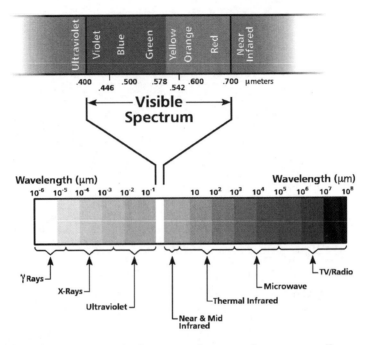

Maxwell's work showed that a continuum exists among all types of electromagnetic radiation and that all electromagnetic radiation has a common mathematical description and origin. Typically engineers divide the continuum into segments based on the radiation's wavelength because the useful behavior of the waves depends on its wavelength. The most important divisions are radio, microwave, infrared, the visible region, ultraviolet, x-rays, and gamma rays. Higher frequencies have shorter wavelengths, and lower frequencies have longer wavelengths.

Eight Amazing Engineering Stories

Design and Operation of the Magnetron

The two key characteristics of a magnetron are: How the magnetron in the oven generates great power, and how it produces oscillations. A magnetron's power comes from the large voltage between its positive and negative electrodes and the oscillations (the RF) from cavities cut into the negative electrode. To understand the magnetron's operation, it's best to look at its operation step-by-step.

First, a magnetron has a negative electrode at the center (called the cathode) surrounded by a ring-shaped positiveelectrode (anode). When a high voltage is applied between the two electrodes, current flows because electrons zip from the negative

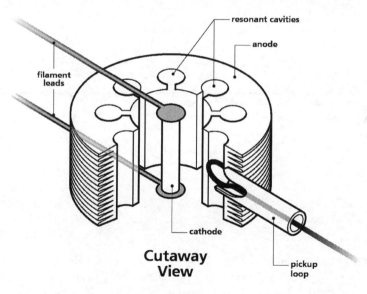

The essential parts of a magnetron are: 1) the cathode that produces electrons, 2) the anode (typically copper) that captures electrons, and 3) the resonant cavities that create electromagnetic waves.

to the positive electrode. This voltage heats the cathode so that electrons "boil" off and fly to the copper anode.

Second, a magnetic field is applied perpendicular to the path of the current. Typically, this comes from a permanent magnet built into the magnetron. An electron traveling in a magnetic field will feel a force, and the applied magnetic field will cause the electron to curve in a plane perpendicular to the magnetic field.

For a magnetron, this plane would be a circular slice when viewed looking down the electrode. As the magnetic field strength increases, the electron's path will curve more and more in the interaction space between the two electrodes. This creates a cloud of rotating electrons that in turn rotate slowly around the cathode.

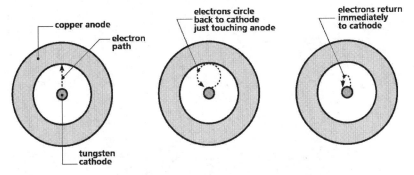

Applying a large negative voltage to the filament causes it to heat and emit electrons. Left *If there is no magnetic field, the electrons radiate along a straight line to the copper anode. (Only one electron path is shown for clarity, but it would look like the spokes of a bicycle wheel.)* Middle *If we turn on a modest magnetic field perpendicular to the electron path, the electron will circle back to the electrode just grazing the anode.* Right *A strong magnetic field will cause the electron to immediately return to the cathode without coming near the anode's inner wall.*

It's important to note that at this point, no radio frequency waves come out of the magnetron, and all we have is a series of rotating electrons. To get waves, we need to add resonant cavities.

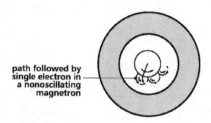

If the anode has no cavities carved in it, an electron will follow a circular path that in turn circles the cathode.

Third, the magnetron is operated with the magnetic field set such that the electrons traveling from the cathode to the anode just brush the edges of the anode. Cut into the anode are circular or triangular cavities with only a narrow opening to the interaction space. As the orbiting electrons pass by these openings, they create oscillations within the cavities. Think of it like this: When you blow over the top of an empty bottle, it makes a deep sound because you are setting up or creating a standing wave or oscillations in the bottle. The electrons do the same thing as they sweep past the slots, although here the oscillations are in the electromagnetic region of the spectrum, whereas when you blow over a bottle, you are making sound waves.

How Heating Tungsten Produces Electrons

To understand the materials of construction for the magnetron, or why it is built out of tungsten and thorium, let's examine a few critical properties of these elements. Energy supplied to a metal can liberate electrons from the solid's surface into the open atmosphere; the amount of energy needed for this to happen is called the metal's work function. In the microwave, this energy is supplied through the addition of heat, which is known as thermionic emission. Many metals exhibit this effect, but none have the other properties of tungsten that make it essential for the microwave oven's magnetron. Tungsten has the distinction of having the highest melting point of common metals. This is useful in that you can run the magnetron at higher voltages and produce a larger number of electrons than a metal that melts at a lower temperature. Yet, because of tungsten's high work function, pure tungsten would operate at too high a voltage, requiring too much power. To reduce the required power, yet keep the high melting point of tungsten, engineers make microwave oven magnetrons from a combination of tungsten and thorium. Thorium has a much lower work function than tungsten and so will emit electrons at a lower voltage, but on its own would melt at the temperatures that result from the necessary voltages to operate the unit. In a mixture of the two, the thorium can be turned into thorium oxide at the electrode's surface, then as the surface thorium evaporates during operation, it is continuously replenished by diffusion from inside the electrode.

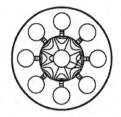 

If the magnetic field is such that the circulating electrons just touch the cavity opening, an oscillating electromagnetic wave will be formed. Left *The lines indicate the electric field created by the cavities. They are shown at one instant in time; these fields reverse direction periodically as the magnetron operates. These oscillating fields influence the path of the electron.* Right *They create a "spinning pinwheel of electrons" when combined with the magnetic field perpendicular to the plane of the paper. (The ends of magnetic field "arrows" are shown as circles in the cavities.) Shown within a pinwheel spoke is the path of a single electron as it spirals toward the anode. This pinwheel rotates, creating an electromagnetic wave with a great deal of power.*

Fourth, once the oscillations from the resonators have been established, the electrons maintain the oscillations that would otherwise die out as the result of joule heat losses in the copper block. A critical step, and the step that gives the magnetron high power, is that the RF fields from the cavities draw energy from the orbiting electrons. In other words, oscillations are maintained at the expense of the energy of the orbiting electrons. In a magnetron with no cavities, the electrons would cycle back toward the cathode (if the magnetic field were strong enough). When this rotating wheel induces oscillations in the cavities, oscillating magnetic and electric fields extend from the cavities. These fields speed up or slow down the electrons. When they speed it up, these electrons just fly back to the cathode because the RF field increases their curvature. In fact, those electrons that absorb energy—those that speed up—from the RF field are eliminated at once. But a number of electrons are slowed down, and their curvature is decreased.

These slowed-down electrons will eventually reach the anode and thus contribute to the current and give some of their energy to the RF field. This results in a distortion of the rotating cylindrical sheath that occurred when no cavities were present; the sheath now appears as a smaller cylinder with four spoke-like ridges running parallel to the axis. This pinwheel rotates with an angular velocity that keeps it in step with the alternating RF charges on the anode segments, and the ends of these spokes may be thought of as brushing the ends of the segments and thus transferring charge from cathode to anode.

Why Do the Cavities Resonate?

It seems counter-intuitive that a simple hole like the cavities would cause electrons to resonate, but the phenomenon can be understood by modeling the system as a grandfather clock.

A pendulum moving back and forth, as in a grandfather clock, is in resonance. This means that it moves back and forth at a specific frequency. Resonance occurs when a system is able to store and easily transfer energy between two or more different storage modes: In the case of the pendulum, this is the change from kinetic energy to potential energy. Resonance isn't limited to mechanical systems; it can also occur in electrical systems by storing energy in magnetic and electric fields. The latter are analogous to the two energy storage modes—kinetic and potential energy—of a grandfather clock.

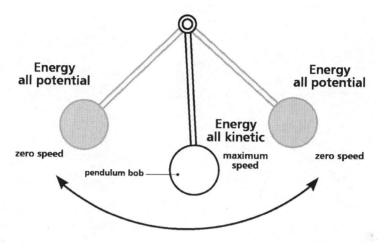

As the pendulum's bob reaches the outer edge of its arc, its speed drops to zero. At this point, its energy is completely potential, but it gives up that potential energy in exchange for kinetic energy as it begins its descent back to the other side of its arc. At the very bottom of the arc, the pendulum is moving at its fastest speed; all its energy has been converted to kinetic energy. It gradually turns this kinetic energy into potential energy as it reaches the limit of its arc on the other side. A key point to note from this example is that systems have specific resonance frequencies; once the pendulum begins to swing, it will swing at only one frequency until its motion dies out because of wind resistance.

In an electrical circuit, the two modes of energy storage that create the oscillation are electrical and magnetic fields. A simple oscillating circuit contains an inductor and a capacitor.

When current flows through an inductor, the inductor stores the energy as a magnetic field, and the capacitor stores energy in a electric field.

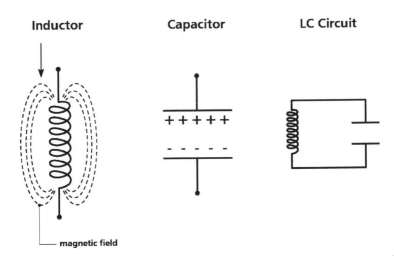

Left *An inductor stores energy in a magnetic field. The energy from a current flowing through the coil is stored in the magnetic field generated by the coils. If it were short-circuited, it would release its stored energy as current.* Middle *A capacitor consists of two separated plates. As current flows, charge builds up on the plates until the capacitor is full. If the capacitor were short-circuited, it would release its energy, creating a current.* Right *If we combine the two elements, we make what's called an LC circuit —L is the symbol used for inductance, and C for capacitance. As described in the next figure, this circuit can oscillate.*

Store is a key word here: It implies a time-varying element. Let current flow through an inductor, and it will store the current's energy in a magnetic field until full. Then, when the current is shut off, the inductor's magnetic field slowly dissipates as current. In a capacitor, something similar happens: When current flows through the capacitor, it stores a charge that is released when the current stops. At its simplest, an inductor is just a tightly wound coil of wire. As we've learned, current flow creates a magnetic field. The coil of the inductor concentrates this magnetic field. A capacitor consists of two metal plates held apart a short distance.

When current flows, charge builds up on each plate and creates an electric field. If we place an inductor and a capacitor in a circuit and charge only the capacitor, this circuit will resonate.

You might ask whether such a resonant circuit can be used to generate radio frequency waves, since all we need to generate electromagnetic waves are oscillating magnetic and electric fields. Indeed, this does work; in fact, the earliest radios used circuits comprised of capacitors and inductors modified by a vacuum tube or transistor to keep the circuit oscillating. The latter two elements did the electronic equivalent of mechanically tapping the pendulum of a clock when its oscillations began to decay. So, it would seem that this would be a great way to make RF in the microwave range, and it does work to a degree. We can produce radio frequencies but eventually, at very high frequencies, the rapid back-and-forth motion burns out the circuit elements.

For high-frequency radio waves in the microwave regions, we can build the circuit into the copper anode without any wires or apparent circuit elements. A cavity alone does the trick. In the anode, engineers cut cylindrical cavities. Viewed from the top, it looks like a loop of wire where the ends are very narrow. If current flows around the loop, the cavity causes the cylinder to behave like an LC circuit. The sides of the narrow slot are so close together that they act like a capacitor—it stores energy as charge builds up. This behavior is completely analogous to the resonator created by blowing over the top of a bottle.

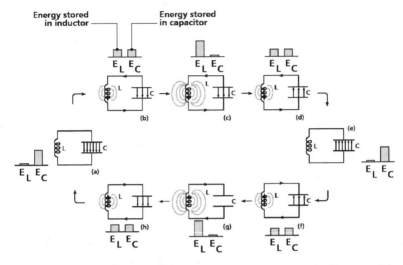

As the capacitor discharges, its current flows into the inductor. Because of the flowing current, the inductor generates a magnetic field that stores the energy contained in the current. Eventually, the capacitor becomes completely discharged. Current no longer flows, so that inductor begins to release its energy as current flows, which causes the capacitor to become charged. Once the inductor is completely discharged, current no longer flows, and the capacitor then releases its energy, starting the cycle over again. Shown underneath every step is the energy stored in the inductor or capacitor. The total energy moves back and forth between the inductor and capacitor to store the energy, just as a pendulum's energy oscillates from potential to kinetic. The rate at which this exchange happens in the circuit (the resonant frequency) would depend on the size of the inductor and capacitor. Of course, the oscillations cannot go on forever. Just as in a pendulum clock, the resonance decays. The energy being shuttled back and forth is lost as heat in the wire, the inductor, and the capacitor.

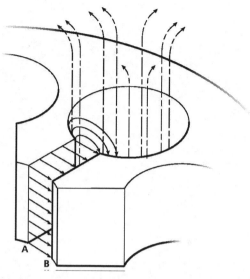

In this cavity, current moves around the circular opening. The gap creates a capacitance. The energy transported by the current gets trapped in the magnetic field created, making the cavity act like an inductor and setting up an oscillation just like in an LC circuit. The solid arrows indicate the electric field, and the dotted arrows show the magnetic field.

Primer: Electrons, Energy Levels, and Light Emission

AN ATOM OR MOLECULE consists of a charged nucleus or nuclei surrounded by a cloud of electrons—a cloud because really we can only talk about the probability of an electron's location.

Diatomic molecule — Nucleus, Nucleus, Shared electrons form bond

Silicon (crystalline solid) — Shared electrons form bonds, Silicon nucleus

Bonds form between atoms when electrons are shared. While we have a commonplace image of an electron as a small sphere orbiting a nucleus, it's best to think of the electron as a cloud of probability. In the molecule on the left, *you can see that the electron is more likely to be found near the nuclei than far away. A crystal forms (as shown on the* right) *when many atoms share electrons to form a lattice. Near the nucleus, the cloud is the most dense, and as we move away from the nucleus, the electron density tapers off.*

Ground states and excited states. Imagine a single electron from that cloud. That electron will be attracted by the positive charge of the nucleus and so it will desire to move toward it, but it will be repelled by the other electrons clustered around the nucleus. Thus, it will find some place to "rest" that balances these two forces.

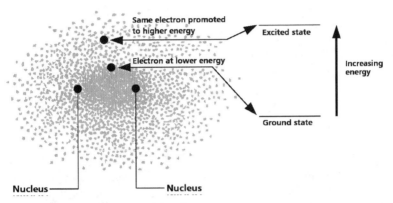

Shown on the left is an electron in its ground state: It sits on the probability cloud at its most likely location. We can add energy to move this electron further away from the nuclei. This is called an excited state or an excited electron. Typically chemists represent these two states on a diagram as shown on the right. These electrons will eventually return to their ground states. We could, of course, move an electron so far from the nuclei that it leaves the molecule forever. That is called ionization.

We could do work and add energy to the electron, and that would move it away slightly from the nucleus. We would call this an excited state of the molecule or say that the electron is excited. Naturally, the electron will eventually return to its lower energy position, called the molecule's ground state, once we stop supplying energy to it. Think of a playground swing: Its natural state—ground state—is just hanging underneath the bar. If we add energy by pushing it, we can create an excited state. With that energy we can do many things, like knock something over in its path. However, if we use the energy to do that, the swing loses its energy quickly. Regardless of what we do, the swing will eventually return to its ground state: All the energy will be lost as heat, either in the air resistance of the swing, or in friction between the swing's chain and the metal support bar. Like a swing in motion, there are many ways an excited electron can return to its ground state, but the electron has an odd property not apparent in the swing's motion.

Quantization. You would think that an electron could occupy any value of energy around the atom. You might guess that we could add whatever energy we want and move the electron any distance from the nucleus, but we cannot.

Let's return to our familiar example of a grandfather clock. When a pendulum moves back and forth, it sweeps continuously through every section on its arc; it moves smoothly without jerks, changing its energy slightly on each part of its arc. Yet that motion is only an illusion; in reality the pendulum moves in incredibly small, piecewise steps, jumping from one position to the next, never existing in between. We call this motion *quantized*: a series of discrete places—or, more precisely, energy levels—where the pendulum can be. To our eye, the pendulum's motion appears continuous because the difference between the energy levels is far too small for the human eye to detect.

According to quantum mechanics, the pendulum can only have energy values—and thus locations on its arc—of: $(1/2)h\nu$, $(3/2)h\nu$, $(5/2)h\nu$ and so on. Here h is Planck's constant and ν is the frequency of the pendulum, typically 2 cycles per second for a grandfather clock. Planck's constant is tiny: 6.626×10^{-34} joule-seconds, where one joule of energy is about what you'd feel if you dropped this book on your head from about a foot above your head. The tiny value of Planck's constant means the spacing between energy levels associated with the possible locations of the pendulum are very small. In our everyday world, the energy levels associated with the pendulum are indistinguishable from one another, but when we get down to the size of atoms, these effects become important.

At the microscopic level, then, only discrete and quantized energy levels are available, hence the terms "quantum" and "quantum physics." In the drawing of energy levels in the figure above, there is nothing in between the two levels. This isn't just a peculiar phenomenon of electrons in atoms; at its core, our world is completely quantized, even in terms of motion.

Why is the world quantized? For the purpose of this book, we will accept the following fact: At a small enough scale there are discrete energy levels to entities such as electrons. We won't bother with anything beyond this, we'll only ask the technological importance of it.

Decay of excited electrons. We can excite electrons by adding energy—heating an object until it glows as in heating a filament, passing electricity through it like a fluorescent lamp, injecting current as in an LED (light emitting diode), or using the energy from a bright light, as in a ruby laser. All these ways to add energy are analogous to pushing a swing to "excite" it. An excited electron decays, losing energy by returning to its ground state, in several ways. It can do so in a way that emits no light, called non-radiative decay, where instead of light, it gives off heat. To be specific, it can release energy by vibrating the molecule or by inducing a collision or rotation. The electrons returning to lower energy could also cause the molecule to dissociate or break up.

Lastly, the excited electron could cause the molecule to photochemically react. But most important technologically is a radiative loss of energy; that is, the returning energy being released as a burst of electromagnetic radiation.

There are many ways to induce this burst. Of main importance for a laser is using light or electricity to create the excited states. If we use current, as in an LED, it's call electroluminescent. There are two ways that an electron can give off light: spontaneous emission, which results in incoherent light, or stimulated emission, where many electrons radiate together, creating light that is in phase. The former occurs in light bulb filaments and glow-in-the-dark toys, the latter happens in a laser.

Chromium, Helium, & Neon
[kroh-mi-uhm] | Symbol: Cr
    Atomic number 24 | Atomic weight 51.9961 amu
[he-lee-uhm] | Symbol: He
    Atomic number 2 | Atomic weight 4.002602 amu
[nee-on] | Symbol: Ne
    Atomic number 10 | Atomic weight 20.1797 amu

In this chapter, we highlight the role of elements used in the creation of laser light. The first laser, which used chromium, appeared in 1960, followed rapidly by those based on helium and neon.

The word chromium comes from the Greek chromos, *meaning color. Indeed, the oxide of this silvery white-gray metal, discovered in a Siberian gold mine in 1766, creates vivid colors in a variety of compounds. For example, the mineral* alexandrite *contains a small amount of chromium which absorbs light over a narrow range of wavelengths in the yellow region of the spectrum. Alexandrite from Russia's Ural Mountains (it is named after the Russian Tsar Alexander) looks green in daylight but red by incandescent* light. Ruby is largely a crystal of aluminum and oxygen, where chromium replaces a small fraction of the aluminum atoms. The chromium strongly absorbs yellow-green light, which makes the gemstone glow red when a blue light shines on it. This red glow forms the basis of a ruby laser.

The inert duo of helium and neon revolutionized lasers. Both are colorless, tasteless, and odorless gases. Neither reacts with anything strongly; in fact, neon won't even form a compound with itself, a rare feat for gases. Helium's discoverer detected it in the sun long before anyone found it on the earth. Using a prism, French astronomer Jules Janssen observed a yellow band in the light coming from the sun. He surmised, correctly, that the line corresponded to a new element. British astronomer Norman Lockyer discovered helium at the same time, and he named it in honor of the sun, basing it on helios, the Greek word for sun. Stars, such as the sun, produce huge amounts of helium by the fusion of hydrogen. No surprise, then, that it's the second most common element in the universe after hydrogen (the two together account for 99% of all observed matter).

In contrast to helium's extraterrestrial discovery, neon became widely available for the most human and *earthbound of reasons: war. At the start of the nineteenth century, a Frenchman, Georges Claude, searched for a way to recover oxygen from air. What really ignited demand for oxygen was an arms race: The steel industries of Europe needed lots of pure oxygen to create extremely strong steel for armaments. By freezing air, Claude was able to extract pure oxygen, but since air contains more than just oxygen, he also recovered the rare gas neon. Neon gas was so expensive at the time that, prior to Claude's work, it was used only for exotic purposes. For example, in 1897, a special light display was made using neon and other rare gases to celebrate Queen Victoria's Diamond Jubilee. Today, we see it all the time: The glow inside these neon lights is a sort of tamed lightning used in displays everywhere.*

Electricity strips electrons from neon, and the electrons then smash back into the positively-charged remains, producing a brilliant light. In combination with helium, neon produces one of the world's most useful and inexpensive lasers.

How a Laser Works

O**N JULY** 21, 1969, astronauts Neil Armstrong and Buzz Aldrin walked on the moon. Although getting to the moon was a "giant leap for mankind," few know that one of their tasks for this pioneering mission was to install a small metal plate covered with 100 reflectors made of fused quartz on the surface of the moon. Once the reflectors were placed there, geo-scientists made highly accurate measurements of the separation of the moon and the earth. From these measurements, they learned fundamental facts about how the moon behaves; they learned that it spirals away from earth at a rate of 38 mm a year, that the moon has a liquid core that is 20% of the moon's radius, and they even showed that Einstein's Theory of Relativity predicts the moon's orbit with incredible accuracy. They measured the moon-earth distance by firing a high-powered laser towards the reflector and measuring the time it took the beam to travel there and back. Using the speed of light, they calculated the distance between the earth and the moon to an accuracy of 2 cm, an astonishing accuracy considering the moon is 385,000 km from earth. The amazing part of this story isn't only humans walking on the moon for the first time or the accuracy of the measurement, but that we could, from earth, shine light onto the moon!

This exact measurement was made possible by the invention of laser light. Laser—an acronym for Light Amplification by Stimulated Emission of Radiation—produces light from a phenomenon known as stimulated emission, which we'll explain later in this chapter. Although lasers are commonplace today, they are different from normal light in a few ways. Imagine using a flashlight to illuminate the moon: Its beam would dissipate into the atmosphere long before it reached its destination. Though powerful, the flashlight lacks the three key characteristics of a laser that allow light to make it all the way to the moon and back and still be measurable.

Narrow beam. A key characteristic of laser light is an incredibly focused, narrow beam of light. When the light travels to the moon, the beam spreads out to a diameter of 7 km, but without beginning as an incredibly narrow beam, it would diffuse completely in the earth's atmosphere.

Great intensity. Since all of the light's energy is packed into a narrow beam, it is bright enough to reach the moon, be reflected, and return to earth.

Monochromatic beam. Its single pure color allows it to be detected when it returns to earth: Scientists can separate the return signal by narrowing their search for the return beam to a nearly single color of light.

Using a laser to reach the moon is an exotic, even esoteric, use of a laser, but it highlights how the union of these three characteristics gives lasers immense technological importance. Lasers flash messages through fiber optic lines, read grocery bar codes, and even heal people. Eye surgery, for example, highlights again the importance of all three of a laser's unique properties. Surgeons use green lasers because that color passes through the eye's vitreous humor without being absorbed and thus does not cause damage. The laser beam then strikes the retina where its great power can "weld" a detached retina back in place. The beam's narrowness lets the surgeon affect only the area of the retina that needs to be repaired. Without the three unifying characteristics of lasers, doctors wouldn't be able to perform such surgery.

Key Characteristics of a Laser Beam

By comparing a laser's light to that of a light bulb, we can refine the characteristics that make a laser unique.

Engineers typically recast the three key characteristics of a laser mentioned above in more technical terms. Specifically, unlike incandescent light, a beam of laser light has:

Temporal coherence. All of the peaks and troughs in the waves of light line up. This gives the light beam an astonishingly pure color, best characterized as almost a single wavelength of light. An incandescent light shines with a variety of wavelengths.

Spatial coherence. All of the light rays are evenly and equally spaced. This gives the beam great power: In an incandescent light, the intensity varies—in some places its rays are packed tightly, elsewhere very spaced out.

Collimation. All of the rays from the laser are parallel. This means the beam contains only a small area, giving the user tight control in, say, removing a tattoo or reading a bar code. In an incandescent light, the light rays scatter in all directions, resulting in a beam of overall lower intensity.

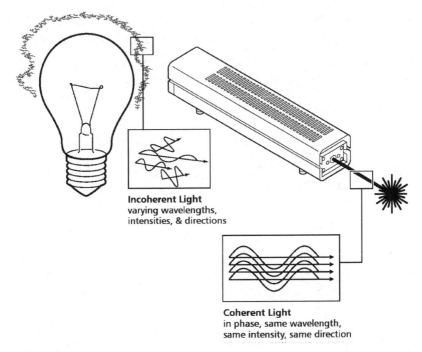

A light bulb produces incoherent light: it is made of many wavelengths, travels in all directions from the bulb's filament, and varies in intensity. Laser light is characterized by coherence—meaning that the light exiting the laser is in phase—of a very narrow band of wavelengths and travels in the same direction.

How does a laser create this amazing beam of light? For that matter, how does anything create light?

Why Does Anything Give off Light?

The basis of all human-made generation of light lies in the movement of electrons within an atom or molecule. If we supply energy—an electrical current, heating, or even a flash of powerful light—we can promote electrons in a molecule from their normal low-energy state to higher energy. Typically, an electron persists in this excited state for only a short time before it returns to lower energy. When it returns, it must give off that extra energy. Most often it just dissipates energy as heat, but sometimes it releases that energy as a burst of light.

In a typical incandescent light bulb, about 90% of the energy supplied dissipates as heat, rather than as visible light. A current passing through a tungsten filament heats the tungsten, promoting some of the filament's electrons to higher energy. As the electrons return to lower energy, some give off light, although most dissipate this energy as heat (think of how hot a light bulb gets after it has been on for awhile). However, heat isn't the only way to "excite" electrons. For instance, if you shine a blue laser pointer on a piece of ruby, it'll glow a robust red. It's absorbing the blue light, using its energy to promote electrons to higher energy, which then decay, releasing a brilliant red color. A similar phenomenon occurs with a glow-in-the-dark football. After being held in the sunlight for half an hour, it'll glow brightly in a dark room. The zinc sulfate copper compound embedded in the football absorbs the higher energy sunlight and then slowly releases it as an eerie green. The sparkle from an irradiated ruby and the glow of the football lie at the core of how a laser works.

Lasers & Spontaneous Emission

In some ways, a laser is like a glow-in-the-dark football on steroids. The key difference in how sparkling rubies and glowing footballs make light and how a laser makes light lies in the number of excited electrons. The difference isn't how it's done, but rather how many times it happens. In a laser, an intense burst of light creates many excited electrons; their decay gives the laser beam its unique properties.

To make a laser, we need a medium with at least three energy levels: a ground energy level and two higher energy levels. With a burst of energy, we can create a population inversion; that is, more electrons in the highest energy level than in the energy level just below it. This inversion initiates a most amazing thing: When one of the electrons returns to a lower energy state (releasing its energy as light), that light ray triggers other excited electrons to release their energy as light—and here's the fascinating part—in phase with the first ray of light released. This stimulated emission gives the laser beam its great coherence. Contrast this with the emission of light by the football: There, the excited electrons decay by spontaneous emission, giving off light rays that are not in phase; to use technical language they have no temporal coherence. By creating the population inversion, we dramatically increase the chance that stimulated, rather than spontaneous, emission will occur.

To consistently and reliably create the stimulated emission at the core of a laser, to put it into practice, requires a great deal of engineering.

Spontaneous emission

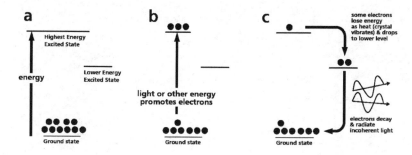

Stimulated emission

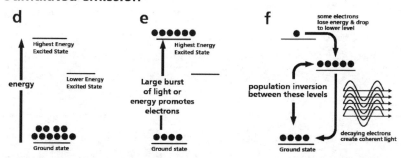

In a glow-in-the-dark toy, the electrons start a) in the lowest energy level, called the ground state, then b) light promotes or excites a few of the electrons to a higher level. These excited electrons drop c) to an energy level between the excited and ground states. The electrons from that level can return to the ground state, giving off light. This spontaneous emission creates incoherent light. In a laser, the electrons also start d) in the ground state, then e) a huge burst of energy rapidly promotes many electrons to the excited state. The excited state electrons drop to an energy level between the ground state and the excited state. When these electrons f) return to the ground state, they give off coherent light because of the population inversion; that is, there are more electrons excited than in the ground state.

Why Does Stimulated Emission Occur?

The laser requires quantum mechanics for a complete description because spontaneous and stimulated emission have no true classical analogies. To deeply understand laser technology requires the use of QED (quantum electrodynamics), which is a step beyond the quantum taught to most engineers. Stimulated emission seems like a very odd concept, yet it is the spontaneous emission of light (light that isn't in phase, like the kind that comes from a glow-in-the-dark football) that truly reveals the strangeness of the microscopic world. Understanding spontaneous and stimulated emission brings the wave-particle duality of light to the forefront of our thinking and highlights its essential "weirdness" in relation to our macroscopic experience of the world. At a very simple level, one can visualize stimulated emission as either a particle or a wave phenomenon.

A photon, behaving like a particle, ricochets back and forth throughout the laser cavity. Like a billiard ball, it hits electrons in their excited state, inducing them to return to a lower energy level and release some of their energy as photons. In that collision, the impacting photon loses no energy—it's an elastic collision similar to what happens to idealized billiard balls. The photon released in turn collides with another excited electron to release more photons that eventually create the great cascade that gives the laser its powerful light. An alternative description to this cascade can be given by viewing stimulated emission as arising from the wave nature of light.

Picture the light racing back and forth within in the laser as a wave. The excited electrons act like antennas detecting the fluctuations of the electric field, which push and pull until the electrons release a wave of light with a phase that is in complete sympathy with the first wave of light.

Although greatly simplified, these two descriptions make sense in relation to our own experience of the world and help us understand how lasers work. Spontaneous emission, though, is much harder to square with our experience of the world.

Spontaneous emission occurs even if there are no photons or light waves (choose your poison) to entice a light from an excited electron. To understand why spontaneous emission happens, we need to appreciate the fundamental way in which the quantum world differs from our human-sized world. Spontaneous emission occurs because on a small enough level, the world is never completely still. Werner Heisenberg showed in 1927 that the position and energy of an object can never be exactly defined at the exact same time. This means that nothing ever stops moving, a notion foreign to our commonsense experience. For example, we think of a vacuum as the absence of everything, yet microscopically it is a hotbed of activity. The vacuum contains a small amount of energy, called zero-point energy, that can induce spontaneous emission. If you find that hard to wrap your head around, keep in mind that there is no analogy from the human-sized world to guide you, so it really makes no classical sense.

The Ruby Laser

Ted Maiman, an engineer at Hughes Research Laboratories, built the first laser in 1960. He knew of the MASER, which used stimulated emission to produce microwave radiation, but he wanted to use the same phenomenon to produce visible light.

Maiman created the population inversion by using ruby as the core of his laser and a flash lamp to supply the energy; the lamp was used in aerial photography and was so powerful it could ignite steel wool. Inside this helix-shaped lamp he placed a cylinder of ruby 1 cm in diameter and 2 cm long.

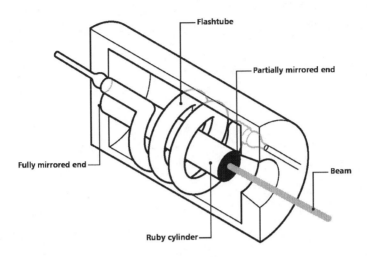

The first ruby laser was only a few centimeters long. Its center was a ruby cylinder mirrored at both ends; one of them, though, was partly silvered so eventually light could escape. A high-powered flash lamp surrounding the cylinder provided the energy to create a population inversion.

He coated the ends of the cylinder with silver to make them mirror-like, leaving a small hole in the center of one mirrored end to allow a small amount of light to escape. Maiman flashed the lamp for a few milliseconds; the intense burst of light caused electrons in the ruby to become excited, although most of the flash was lost as heat. The ruby rod emitted bursts of incoherent red light via spontaneous emission. He kept flashing the light, effectively "pumping" the electrons, and eventually he created a population inversion. The radiative decay of an excited electron then triggered stimulated emission, releasing an avalanche of light with all of its waves in phase. The light swept back and forth across the ruby cylinder, reflected by the mirrored ends. These repeated reflections both narrowed the range of wavelengths of light emitted from the ruby and eliminated any rays not parallel with the ruby cylinder's axis. This creates the highly coherent, monochromatic beam characteristic of a laser.

Maiman succeeded in creating the world's first laser and so was twice nominated for the Nobel Prize. He never won; instead the Swedish Committee awarded it to the inventors of the MASER, which paved the way for lasers.

Maiman's laser had several imperfections that made it non-ideal for portable, everyday use. For instance, it required a lot of energy and gave off large amounts of heat. Ruby lasers could only be easily used in pulsed-mode, rather than continuous, because of the intense heat built up. To make the laser a less-specialized, everyday tool required other lasing media than ruby. Of greatest importance are gas and semiconductor lasers.

Helium-Neon Lasers. By replacing ruby with a gas mixture, engineers create lasers than can be operated continuously simply by removing heat using passive air cooling at low powers or, for high-powered operation, by using forced air from fans. A stellar example of such a laser is the helium-neon laser, which produces very little waste heat, is easy and cheap to construct, and is very reliable. A simple flick of a switch, and a red beam of wavelength 632.8 nm emerges.

At the core of the helium-neon laser lies a sealed glass tube filled with helium and neon with reflecting ends, as in the ruby cylinder. Its inventors replaced the cumbersome flash lamp of the ruby laser with a large jolt of electricity: zapping the gases with 9,000 volts, creating excited electrons in the helium gas. These helium molecules in turn collide with neon gas molecules. The energy levels of the exited helium overlap with the exited states of neon, so the collisions transfer energy to the neon, creating excited electrons in the neon. The stimulated emission of these electrons radiatively decaying gives off lights of wavelength 632.8 nm (a red light) and infrared light of wavelengths 1150 nm and 3390 nm that human eyes cannot detect.

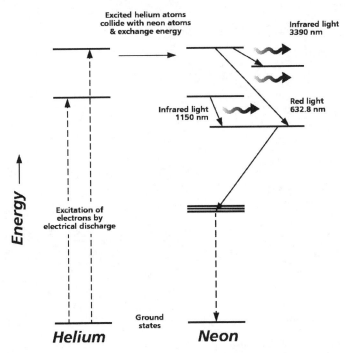

A helium-neon laser contains a mixture of both gases. A voltage discharge excites the electrons in the helium atoms that, through collisions, transfer their energy to the neon atoms. (Helium and neon were chosen because their energy levels in the excited state match up closely.) The electrons in the neon decay to give off light. There are three main transitions, two in the infrared and one in the visible (red) part of the spectrum.

> **Why Use Two Gases in a Helium-Neon Laser?**
> Neither gas alone would make a very good laser. The excited helium atoms are very stable in the sense that the electrons stay excited for a long time but cannot easily radiatively decay. In contrast, the neon won't stay excited for long, but it does radiate easily. The combination of the two uses the best properties from both gases.

A gas laser has the huge advantage over the first ruby laser in that it doesn't produce as much heat. The radiative decay of the excited neon atoms is from one higher-energy state to another, rather from an excited state to the ground state as in a ruby. A helium-neon laser can therefore operate at a lower temperature than a ruby laser can. Recall that at the quantum level, the temperature of a substance reflects how the electrons are distributed among a molecule's energy levels: A higher temperature means more electrons in higher energy states. So, in a three-level system—that is, three energy levels—like ruby, we need to promote a large number of electrons from the ground state to a higher energy state to get a population inversion. (The key population inversion is between the ground state and some higher energy level.) In a four-state system like helium-neon, we only need to get a population inversion between two higher energy states. The radiative decay occurs when an electron decays from one higher energy state to another high energy state, rather than to a ground state. This means that we can have more electrons in the ground state for helium-neon than for ruby and thus operate at a lower temperature. You might also think of it this way: We need to promote more electrons in a three-level system than in a four-level system. Each of the excited electrons generates heat—they make the crystal vibrate—as it drops to a lower level. More electrons thus means more heat.

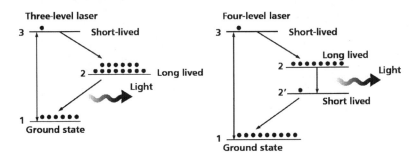

To achieve a population inversion in a three-level system, more electrons need to exist in energy level 2 than in the ground state, energy level 1. In a four-level system, the population inversion necessary for stimulated emission needs to exist only between levels 2 and 2'. This means that we don't need to promote as many electrons in a four-level system (like HeNe) than we do in a three-level system (like ruby). This is what allows a HeNe laser to operate cooler than a ruby laser.

Who Really Invented the Laser?

The headline on a 1965 *New York Times* story announced—on page 96—a possible problem for the rapidly growing laser industry: "A Laser Patent Fought in Court: Former Graduate Student Seeks Rights to Device." At stake was millions in licensing revenue. The plaintiff, Gordon Gould, claimed to have invented the laser while working as a graduate student at Columbia University; he had failed to get his doctorate there because of his preoccupation with laser research. He even noted that he and his wife separated because of his obsession with lasers. Gould's lawyer claimed that Gould conceived the laser in his Bronx apartment in 1957, and then had a notary public in a nearby candy store notarize his notes and drawings. Gould didn't file for a patent, though, because he thought you had to have a working model.

At nearly the same time Charles H. Townes, a professor at Columbia, and his post-doctoral researcher Arthur L. Schawlow developed the same idea. They filed for a patent in July of 1958. At issue was the parallel mirrors that cap the ends of a resonant cavity—a key, perhaps *the* key, to operating a laser. Eventually Townes and Schawlow secured a patent—and separate Nobel Prizes—for their work. Gould received a relatively unimportant patent for laser production of x-rays in 1968. The 1965 lawsuit, though, started Gould on a three decade quest to gain priority for inventing the laser.

He fought and fought until by 1977 he finally received a patent for the optically pumped laser amplifier; that is, the heart of every commercial laser! Soon thereafter Gould was also awarded nearly 50 patents, which covered most of the laser design used commercially. Armed with this portfolio of patents Gould teamed with financial backers to campaign for all manufacturers to pay royalties on nearly all lasers. Even though Gould gave away 80% of his profits to his backers, he still earned, according to Nick Taylor, an expert on the history of the laser, "millions upon millions of dollars."

Resonator Cavity

While the light generated by the ruby or the helium-neon gas (otherwise known as the gain medium) seems the most exciting part of a laser, the real key to making the precise wavelength as well as the great collimation of laser light lies in the design of its cavity. When a laser first begins operation, the wavelength of light generated in the gain medium is not very sharp by laser standards. For example, when a decaying electron in neon emits light in a HeNe laser, that light consists of wavelengths ranging from roughly 620 nm to 640 nm. While this difference of 20 nm would be indistinguishable to our eye, it would destroy the usefulness of a laser. For a typical HeNe laser, the range of wavelength is an amazing 0.01 nm—from 632.79 to 632.8 nm! Without the cavity, not only would the range of colors be broad by laser standards, but the light would not be highly collimated; that is, its rays would go in all directions. The resonator cavity, the area between the mirrored ends of the ruby cylinder or tube containing helium and neon, both narrows the range of wavelengths and collimates the beam by causing the light's rays not parallel to the cavity's axis to destructively interfere and die out.

The repeated reflections inside the cavity create a standing wave. (See the primer on waves that proceeds a discussion of the the microwave oven for background on standing waves.) This means that only light of a particular wavelength can exist in the cavity. Specifically, only light with a wavelength such that its "nodes"—where a waveform passes the x-axis—have a value of zero at the mirrored ends of the cavity, or light whose half-wavelength is an integral multiple of the cavity's length. An infinite number of these wavelengths can exist in a cavity, all spaced equally in energy. So, if we choose the cavity size correctly, only one or a few of these allowed wavelengths will exist. When these allowed wavelengths are convoluted with the slightly broader range of wavelengths emitted by the radiative decay of the gain medium, the result is a narrow single wavelength.

Eight Amazing Engineering Stories

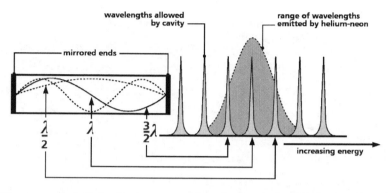

Shown on the left is the resonant cavity of a laser. An infinite number of wavelengths can exist inside the cavity as long as they are an integral number of half-wavelengths. Shown above are three of those wavelengths: λ, $\lambda/2$, and $3/2\lambda$. (Note that these lines are spaced equally in energy). This spacing depends only on the cavity size and not the wavelength of the laser's light. The right shows the light output from the helium-neon red line: Its width indicates a range of color or, better put, wavelengths. The cavity only allows the waves shown to exit. For this 5 cm HeNe tube—much shorter than a real tube—three narrow laser lines would appear: Those at λ, $\lambda/2$, and $3/2\lambda$, with the λ wavelength the most intense. (Here the wavelengths are drawn schematically for clarity.)

In engineering such a cavity, the most critical part is the degree to which the mirrors are parallel. If they deviate by more than a few percent, the cavity will allow all sorts of wavelengths to exist within the cavity and destroy the monochromaticity of the laser. In practice, this means the mirrors cannot deviate from being parallel by more than one-third of the operating wavelength. For the first ruby laser, Maiman built his mirrors so that at no point did the distance between them differ by more than 694.3 nm/3 or 231.43 nm from top to bottom.

Semiconductor Lasers

Although ruby lasers proved that a functional laser is possible, and HeNe lasers illustrated their practical usefulness, of greatest importance in our world today is the semiconductor laser. They're used in laser pointers, bar code readers at grocery stores, and in laser printers. Their key advantage over solid-state or gas lasers lies in the fact that they are compact and cheap because they can be made with the automated methods used to manufacture computer chips.

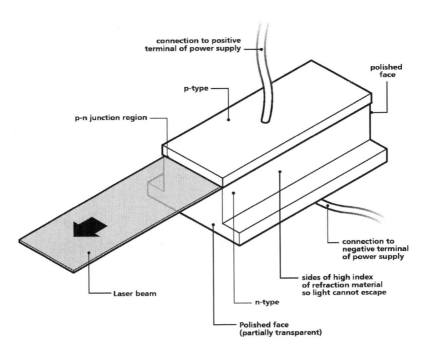

A typical simple semiconductor laser. Light is made at the junction between an n-*type and a* p-*type semiconductor. The sides of the laser are of high refractive index so no light escapes; the ends are highly mirrored to create a resonant cavity.*

Most semiconductor lasers contain the same basic components as the ruby or helium-neon laser, although the method of producing light comes from electroluminescence. That is, a current drives the stimulated emission, not a flash lamp or voltage discharge. A junction of two different types of semiconductors replaces the ruby or helium-neon mixture. Two edges of this "sandwich" are surrounded by material of a higher index of refraction, so no light can exit the sides. The other two ends are highly polished to create the familiar resonate cavity. When current is injected into the junction, light appears as a colored band of light emitted from the area.

The two materials making up the junction are a p-type and n-type semiconductor. p-types transfer current by positive carriers called holes, and n-types by negative carriers called electrons. (See the end of the chapter for a section on semiconductors.) The junction of these two semiconductor types creates a diode; that is, the device allows current from a battery to flow in only one direction. The transfer of current occurs at the junction between the two types of semiconductors: When the battery injects electrons in the n-type semiconductor, the diode allows electrons to flow out at the junction: The electrons recombine or neutralize with positive holes. When they recombine, they can give off light.

The energy difference between a separated electron and hole and their combination is analogous to the discrete energy levels in a ruby or helium-neon laser. The energy of the separated pair is higher than when they are together; therefore, when they combine, they release energy. Just as in a ruby or a mixture of helium-neon, the energy can be released as heat—the vibration of the semiconductor crystal—or it can give off energy as a burst of light. Rather than create a population inversion by a flash lamp, energy is added by passing current through the diode. By "injecting" a large amount of current flowing through the diode, we create many electrons that can then combine with holes, giving off light and inducing stimulated emission. The edges of the semiconductor sandwich are highly polished to collimate the laser.

Erbium Amplifier for Fiber Optics Cables

Understanding stimulated emission allows us to see how one of the cleverest engineered objects of all time works. It's called an erbium amplifier, which is used in fiber optic cables. To understand the revolutionary nature of these devices you must understand something about sending signals through fiber optic cables.

A signal is encoded as flashes of light that are then sent into a fiber optic cable. Via total internal reflection they "bounce" down the cable. While cables do a great job of transmitting the light pulses, the light eventually escapes out the walls of the fiber. Engineers call this the attenuation of the signal; typically a signal will travel many miles until it has dropped to about 1% of its original strength. At this point it needs to be amplified or it will be lost forever. For example, TAT-8, the first undersea cable across the Atlantic Ocean that connected Tuckerton, New Jersey with Widemouth Bay, England, and Penmarch, France, used amplifiers every 30 miles. These repeaters were complex and energy intensive units that used photocells to convert the light to an electrical signal, amplified or boosted that newly transformed signal, and then converted it back to light so it could continue 30 miles to the next repeater. These repeaters were expensive, cumbersome, and, being complex, a source of potential failure. An erbium amplifier, called an EDFA, boosts the light using stimulated emission, eliminating the need for converting the optical signal to an electronic one.

Here's how the amplifier works. Erbium has a rich electronic structure; that is, it has many energy levels that will emit light once its electrons are excited. Inside the glass fiber optic cable engineers build sections, every 30 miles or so, that are "doped" with erbium. Using an infrared laser, they can populate the higher energy levels of erbium and create a population inversion. When the optical signal arrives, its photons induce these excited electrons to decay—recall that this is stimulated emission—and this burst of light adds to the incoming signal thus boosting it optically. The emission is stimulated so the photons produced are exactly in the same phase and direction as the amplified signal, and, importantly, the signal is amplified along its direction of travel only. The greatest advantage of these erbium-doped amplifiers over electronic amplifiers lies in multiplexed signals.

To increase the capacity of a fiber optic cable, engineers send multiple messages simultaneously. They do this by encoding each signal as a different wavelength of light. (Essentially each signal is a different color, but the wavelengths are so close together that our eyes could not tell the difference.) Picture these signals traveling down the the fiber optic cable, slowly losing intensity until they need to be amplified. As they enter the erbium enriched section they are automatically boosted individually. The incoming light of a particular wavelength matches exactly one of the energy splittings in erbium, another signal matches another wavelength.

Lasers began as the stuff of science fiction. First described in Alexey Tolstoy's prescient 1927 novel *The Hyperboloid of Engineer Garin* as a "filament of light of any thickness desired," the laser was used as a death ray. The laser's exotic and supposedly deadly properties stayed a staple of science fiction TV shows and movies even after its invention. Yet with age, it mellowed to assume the most commonplace of tasks—like quietly, unintrusively scanning groceries.

Eight Amazing Engineering Stories

In Depth: Semiconductors, Electrons & Holes

Recall that there are metals that conduct electricity freely, insulators (like rubber) that don't allow electricity to flow, and semiconductors that conduct electricity but not as well as metals. What's most important about semiconductors is that, unlike metals, semiconductors can conduct electricity by two mechanisms: the movement of negative or positive charge carriers. A metal can only transfer by negative charges. We call a semiconductor that uses mostly positive charge carriers (called holes) a *p*-type, and that which conducts electricity by negative charge carriers (electrons) an *n*-type. These two ways to conduct allow engineers to create a solid material where charge can pass in only one direction.

Engineers make a diode by placing *p* and *n* type semiconductors together. The *n*-type has a number of atoms that have an extra electron, and these electrons are mobile. When the electrons move, they leave a fixed positive charge—keep in mind, though, that the atoms themselves aren't mobile. The *p*-type semiconductor has acceptor atoms (fixed in the lattice) that contribute positive charge carriers called holes. When this junction forms, positive holes are attracted to the negative electrons, which will flow across the *pn* junction. You would think these would continue until the charge was even across both semiconductors, but the atoms fixed in place in the lattice prevent this. When a donor contributes a mobile electron, it leaves a positively-charged atom fixed in place. This means that when the positive holes move into the *n*-type semiconductor, they eventually meet a "wall" of fixed positive charges that repels them. Similarly, on the *p*-type side, each mobile hole leaves behind a negatively-charged atom fixed forever in the lattice. This, of course, creates a wall of negative charges, which stops the mobile electrons from moving further.

What Exactly Is a Hole?

We talk glibly of positive holes in a semiconductor, yet they exist only in the minds of electrical engineers. Holes are a fiction, albeit a clever and useful one. It might seem like a hole is really a positron—the anti-matter counterpart of an electron. It isn't. Positrons exist only for fractions of a second and are generated in huge particle accelerators; holes seemingly, in contrast, flourish in all semiconducting devices. Instead, the notion of holes is a shorthand way to identify one of two different ways electrons can move in a semiconductor and thus conduct charge. If a semiconductor conducts electricity mostly by negative charge carriers (an "electron" in electrical engineering parlance), it is called n-type; if it seems to use positive carriers, nicknamed "holes," it is called p-type.

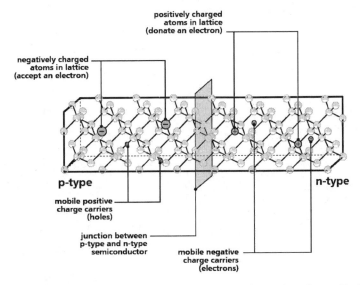

A solid-state diode is made from a p-type and an n-type semiconductor. In the p-type the mobile carriers are thought of as a positive charge called "holes." In the n-type, the mobile charge carriers are negatively charged.

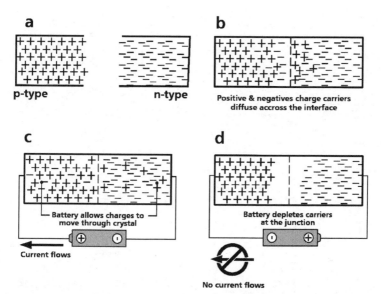

A diode is formed by the junction of p *and* n-*type semiconductors. As shown in (a), we think of the* p-*type as having mobile positive carriers and the* n-*type as having negative charge carriers. When the types are brought together (b), the negative and positive charge carriers diffuse across the interface until an equilibrium is reached. If we hook up a battery (c) such that the negative end is attached to the* n-*type conductor and the positive end to the* p-*type, the negative end of the battery provides a stream of electrons to neutralize the positive charges fixed in the* n-*type semiconductor. This means that the positive holes can flow because the wall of negative fixed charges has been neutralized. Once this has happened, the electrons can flow through the* p-*type. If we reverse the battery (d), a stream of electrons flows toward the* p-*type end. Recall that there was a set of fixed negative charges left from the mobile positive holes, which prevented electrons from moving through the* p-*type. The reversed battery, then, simply worsens the situation, making that end even more negative and thus making it harder for any electrons to move. Thus current comes to a halt.*

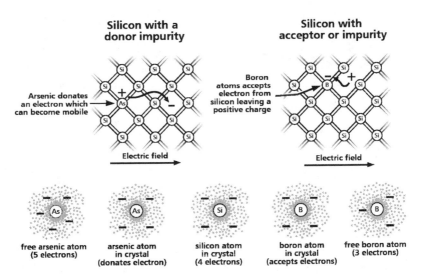

Engineers create n-type semiconductors by adding an atom like arsenic to the silicon lattice. An arsenic atom has five electrons, rather than the four surrounding a silicon atom. This means that arsenic adds an extra electron that can become mobile to a semiconductor. A p-type semiconductor is made by adding an atom like boron, which has three electrons. When added to silicon, boron allows an electron to move from bond to bond. As shown in the next figure, this transfer of electrons can be thought of as a positive "hole" carrying charge.

Engineers can control the type of conduction by replacing some of the silicon atoms with similar atoms. A silicon atom has four electrons that it can share to form bonds. In a pure crystal of silicon, this allows it to bond with four other silicon atoms. If we replace some of these silicon atoms with arsenic atoms—called impurities, which is odd because we want them there!—an extra electron appears. Arsenic has five electrons to share. When we toss it into the silicon lattice, the arsenic shares four of those electrons with neighboring silicon atoms, forming bonds, but the fifth electron is free to roam the crystal. Since the arsenic itself has a charge of +5, it balances that roaming negative charge. Thus the crystal remains electrically neutral over all. At low temperatures, these electrons become bound to the arsenic atoms, but at room temperature, thermal agitation shakes them off and lets them be free to roam. That extra negative charge conducts electricity throughout the crystal. This would be called an *n*-type semiconductor—the "n" stands, of course, for negative.

We could, instead, replace some silicon atoms in pure silicon with an element that has fewer than four electrons. Boron, for example, has only three electrons to share. When substituted for a silicon atom, it cannot complete bonds to all four of the surrounding silicon atoms: It has only three electrons and so forms three bonds. Electrons can jump from one of the neighboring, fully-bonded silicon atoms and completes this bond, leaving of course an electron-deficient bond behind. Now, one could look at this type of motion as an electron moving, although it is different than the electron movement in a *p*-type semiconductor. There an electron wanders from the crystal freely, but here an electron jumps from bond to bond. Now one could look at this as a positive charge carrier that is jumping around the crystal: Every time an electron leaves a bond and completes the fourth bond of the boron, it leaves a positive charge on the bond it left. Most importantly, though, these two mechanisms of charge transfer move in opposite ways to an applied electrical field, just like we would expect positive and negative charge carriers to move.

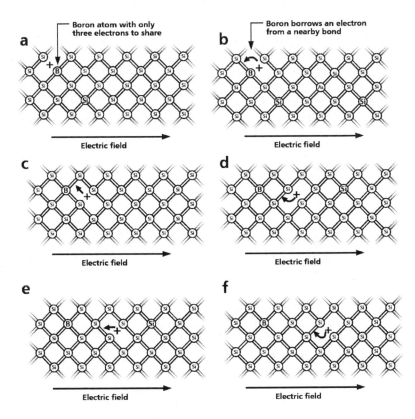

In a semiconductor, all charge is transferred by electrons, but it occurs in two different ways. Shown here is how an electron jumps in a p-type semiconductor from bond to bond when an electric field is applied as indicated. Initially, (a) an impurity boron atom has three complete bonds (two shared electrons) and a bond deficient by one electron. As shown in (b), this impure atom can borrow an electron from a nearby bond. This borrowing can repeat itself across the crystal as shown in (c), (d), (e), and (f). Since the electron-deficient bond has a positive charge, this appears to be a positive charge—called a hole—moving from left to right across the crystal.

NOTE ON THE TYPE

This book is set in Adobe Caslon, a superb digital rendering of the typeface designed by the great English type designer William Caslon (1692-1766). Type designer Carol Twombly (1959-) has captured the unique character and charm of this eighteenth century typeface, used in the Declaration of Independence. Caslon began his career as an engraver of pistols and muskets, eventually moving into letter founding, which made him famous in his lifetime. He produced every letter carefully over a period of twenty years, cutting each by hand. This resulted in idiosyncratic but highly legible type. Some designers take a dim view of this individuality, but despite this professional disdain, the typeface remains one of the best loved.

The typeface Mundo Sans graces the gray boxes throughout the text. A sans-serif typeface created by Carl Crossgrove, it first appeared in 2002, although he started work on it in 1991. He admired humanist sans typefaces like Metro, Formata, Gill, and Syntax. To create Mundo Sans, Crossgrove "used these designs-and, surprisingly, Futura-as models for the proportion, weight, flow, spacing and rhythm." He drew inspiration for the heavier weights from traditional hand-lettered signage, with its heavy sans caps, slightly flaring stems, and humanist skeleton. He succeeded admirably in creating a design clean and distinctive enough for display use while still being understated and suitably proportioned for setting text. It's an understated typeface, but not exactly quiet.

Type designer Zuzana Licko's (1961-) Solex introduces each element. Her typeface brings the industrial sans serif genre into the twenty-first century. Inspiration for Solex came from two sources. First, the slightly condensed, rectangular, grotesque types made for nineteenth century newspaper ads and handbills inspired Licko. And, second, she drew on a typeface called Bauer Topic designed by Paul Renner, who also designed the popular Futura. Renner replaced the Roman widths with mechanical looking, static ones in an attempt to make his type modern and modular.

Made in the USA
Middletown, DE
22 June 2016